FUNDAMENTALS OF STEREOCHEMISTRY AND CONFORMATIONAL ANALYSIS

FUNDAMENTALS OF STEREOCHEMISTRY AND CONFORMATIONAL ANALYSIS

DOC. DR. ING. KAREL BLÁHA, C.SC.

DOC. DR. ING. OTAKAR ČERVINKA DR.SC.

DR. ING. JAN KOVÁŘ, C.SC.

English translation edited by

N. BAGGETT, B.Sc., Ph.D.

Lecturer, Department of Chemistry

University of Birmingham

ILIFFE BOOKS

LONDON

THE BUTTERWORTH GROUP

ENGLAND
Butterworth & Co (Publishers) Ltd
London: 88 Kingsway, W. C. 2B 6AB

AUSTRALIA
Butterworth & Co (Australia) Ltd
Sydney: 20 Loftus Street
Melbourne: 343 Little Collins Street
Brisbane: 240 Queen Street

CANADA
Butterworth & Co (Canada) Ltd
Toronto: 14 Curity Avenue, 374

NEW ZEALAND
Butterworth & Co (New Zealand) Ltd
Wellington: 49/51 Ballance Street
Auckland: 35 High Street

SOUTH AFRICA
Butterworth & Co (South Africa) (Pty) Ltd
Durban: 33/35 Beach Grove

Translated by Ing. Petr Singer
English edition first published in 1971 by
Iliffe Books, an imprint of the Butterworth Group
in co-edition with SNTL—Publishers
of Technical Literature, Prague

Printed in Czechoslovakia

CONTENTS

I. Static stereochemistry

II. Dynamic stereochemistry

PREFACE

About 15 years ago, we organized a series of lectures dealing with stereochemistry at the Institute of Chemical Technology in Prague. The lectures were meant to acquaint the lecturing staff as well as the senior students of the Institute with such results of the rapid development of this scientific field as could not be included in the basic organic chemistry course. In the following years, several similar courses were held for the staff of some research institutes and industrial enterprises, thanks to the initiative of the Czechoslovak Chemical Society. We thus got a chance to convince ourselves of the great interest of chemists in stereochemical problems, and of the need of a brief, informative handbook, which could lead the reader to further individual study.

This book, based on the experience derived from these courses, attempts to be such a handbook, and therefore begins from the knowledge acquired by the graduates of technical schools in the chemical line or by the junior students of the University or Institute, respectively. To a somewhat greater depth, we deal with the problems of optical activity which are traditionally treated in detail at our schools. Therefore, from the teaching point of view, we consider this subject to be the basis of the following chapters, although we are aware that not everybody will agree with us in this respect.

The major part of the book is devoted to the fundamentals of conformational analysis and the stereochemical aspects of the course of reactions, treated sequentially. To a certain extent, we expect the reader to be acquainted with electronic theory and reaction mechanisms, which, of course, could not always be explained in a book unequivocally dedicated to stereochemistry. Considering the brevity of the text and the fact that it is designed for use in practical chemistry, we were unable to treat all problems thoroughly and at length. We therefore concentrated on a consistent and simple classification of the discussed facts, on a more detailed treatment of the chosen examples and on relatively brief theoretical discussion of the fundamental facts. We believe that such an approach is in agreement with the purpose of the book and will furnish the reader with a basis which may eventually be easily supplemented with material from specialized monographs.

K. B.
O. Č.
J. K.

I. STATIC STEREOCHEMISTRY

A. THREE-DIMENSIONAL STRUCTURE OF ORGANIC MOLECULES

Fundamental Principles of Molecular Geometry

The atoms forming organic molecules and ions are interconnected in a variety of ways by means of chemical bonds. In this book we will not discuss the nature of the chemical bond. The fundamentals of this field may be found, for example, in Daudel's book (1) or in various chemistry textbooks. It is sufficient for us to know that the chemical bond is provided by electrons, and that it is exactly defined as regards length and direction in space. The length of the bond, determined as a rule as the distance between the ideal centres of two adjacent, interconnected atoms, is, for the same type, an almost invariable value. On the other hand, the direction of the bond may vary considerably and is determined by the overall structure of the molecule.

Bond Length

Physical measurements, especially spectrometric and X-ray measurements, provide the data for calculations of the length of bonds between different atoms in various groups. The length of the bond is not exclusively dependent on the character of the atoms forming it, but depends also on the immediate surroundings, especially on the number of substituents attached to the atoms. For information, Table I shows the mean values for some of the usual types of covalent bonds. It might be added that the length of the bond may differ if we take it as the shortest straight line between two nuclei or as the length of the line of maximum density of the electronic cloud forming the bond, respectively.

Table I

LENGTH OF BASIC TYPES OF CHEMICAL BONDS

Distance between ideal centres of adjacent atoms (Å). In the table, an atom with four single bonds is designated $_{te}$, an atom with two single and one double bond is designated $_{tr}$, an atom with one single and one triple bond is designated $_{di}$, *i.e.* according to the number of ligand groups.

Bond Type	*Bond Length*
Single bonds	
$C_{te}-C_{te}$	1.536
$C_{te}-C_{tr}$	1.501
$C_{te}-C_{di}$	1.459
$C_{tr}-C_{tr}$	1.47−1.48
$C_{tr}-C_{di}$	1.42−1.46
$C_{di}-C_{di}$	1.387
$N-C_{te}$	1.47
$N-C_{tr}$	1.36
$N-C_{di}$	1.35
$O-C_{te}$	1.43
$O-C_{tr}$	1.34
$F-C_{te}$	1.384
$F-C_{tr}$	1.347
$F-C_{di}$	1.27
$Cl-C_{te}$	1.781
$Cl-C_{tr}$	1.726
$Cl-C_{di}$	1.631
$Si-C_{te}$	1.867
$Si-C_{tr}$	1.853
$Si-C_{di}$	1.850
$H-C_{te}$	1.096
Double bond	
$C=C$	1.335
Triple bond	
$C\equiv C$	1.206

Valence Angle

Not even in a simple chain of atoms are the bonds arranged in a straight line. In an idealized representation by means of lines they form a zigzag line. The atoms attached to one common atom always attempt to occupy positions with the largest possible distance between each other in space. With a constant bond length this condition requires that the four atoms attached to one tetravalent atom occupy the apexes of a regular tetrahedron, in the centre of which the tetravalent atom is located. From this geometrical concept it follows that the angle between any two bonds is the central angle of the regular tetrahedron and is, therefore, approximately 109°. The next atom is attached by means of a bond including the same angle with the previous bond; the new bond need not of course be coplanar with the two preceding bonds. Thus, an atom and its substituents may rotate around the bond connecting it to another atom. We therefore speak of the free rotation of one bond around the other. Later on we will see that the positions which the atoms could occupy as a result of free rotation are not all equivalent and that some arrangements are greatly preferred (see chapter on conformational analysis). Only in relatively few cases does the bulk of the substituents prevent complete rotation around single bonds. Such hindrance results in a special type of isomerism (atropisomerism), which will be discussed later.

Multivalent atoms, and atoms with relatively large radii, *i.e.* with long bonds, generally do not have a bond angle of approximately 109°, like tetravalent atoms of the carbon type, but smaller ones. For example, we imagine a hexavalent atom such as cobalt to have bonds perpendicular to one another, *i.e.* with a bond angle of 90°. The bond angles of small, trivalent or bivalent atoms (nitrogen, oxygen) approach the bond angles of carbon atoms. The spatial concept is such that the atom occupies the centre of a regular tetrahedron. The adjacent, chemically attached atoms are located at two or three apexes of the tetrahedron, and the remaining apex or apexes are occupied by the free electron pairs of the central atom.

Geometrical Isomerism

Geometrical Isomerism at a Double Bond between Two Carbon Atoms

In contrast to the single bond, the double bond is a rigid configuration not permitting free rotation without the outlay of a relatively large amount of energy. Therefore, in the case of the double bond, we have to consider not only the character, number and order of the atoms, but also their mutual distance or relative positions with respect to the so-called double bond plane. This type of isomerism is therefore designated as **geometrical isomerism.**

Whilst a single bond is represented and illustrated as a straight line, it is better to imagine a double bond as distributed within the plane passing through both doubly linked atoms and perpendicular to the bonds linking these atoms with other atoms or groups. This plane divides the substituents attached to the doubly linked atoms into those situated "on the same side" (*cis*) and those located "on opposite sides" (*trans*) of the double bond plane. Let us consider a pair of doubly linked carbon atoms with the ligands a, b, c, d:

a   c
  C=C
b   d

For a formula written down in this way, the "double bond plane" may be represented as a plane perpendicular to the paper and passing through both carbon atoms. All the single bonds attaching substituents a, b, c, and d are then invariably contained in the plane of the paper (Fig. 1).

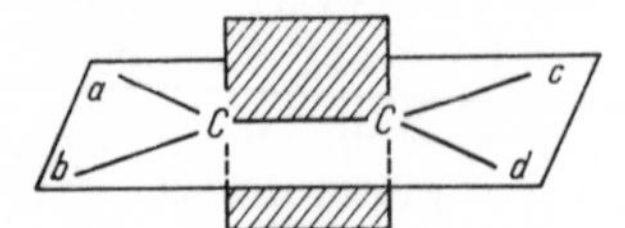

Fig. 1. Geometry of a double bond.
The "double bond plane" is shaded.

The bond angles between the bonds to the same carbon atom are in this case larger than in the case of saturated compounds, and closely approach 120°. We can see that in our formula substituents a and c are situated on the same side of the double bond and are thus in a *cis*-position. However substituents b and d are also situated on the same side of the double bond plane, and consequently are also in a *cis*-position. On the other hand, substituents a and d or b and c, respectively, are on opposite sides of the plane and therefore in *trans*-positions. If at least two substituents on the two atoms of the double bond are identical, we choose them as reference substituents and designate the whole molecule, by means of the respective prefix, according to the common position of these identical substituents. It is obvious that compounds having the same substituents on one atom of the double bond (if a = b) cannot form *cis-trans* isomers. (See NOTE on p. 64.)

Geometrical Isomerism of Cyclic Compounds

We encounter geometrical isomerism in the case of cyclic aliphatic compounds as well. We may imagine that the atoms forming the ring are contained in one plane dividing the substituents—as the double bond plane did—into those located on the same side and those situated on opposite sides. We designate the former *cis*-substituents, the latter *trans*-substituents, in the same manner as with the double bond isomers. If only two substituents are attached to the ring, we may designate the whole molecule according to their common position, by means of a prefix. If there are more substituents, it is better to attach the prefix *cis*- or *trans*- to each pair separately, or to pick one substituent to which the designations of the others will refer. Moreover, a special designation of geometrical isomers has become customary for certain groups of polysubstituted cyclic compounds (*e.g.* for the inositols). As a matter of fact, the ring-forming atoms are not coplanar, with the exception of the smallest rings. However, this has no influence on the existence and designation of geometrical isomers and we may always represent the molecule as a planar figure in order to determine the designation of the isomer. The occurrence of geometrical isomerism is independent of whether the ring is composed of carbon atoms only or of other atoms as well.

The ring may also incorporate a double bond. If the carbon ring is small, three- through seven-membered, all its atoms are bound to be situated on one side of the double bond plane since the total, almost invariable length of the bonds and the only slightly variable bond angles do not permit the connection of doubly linked carbon atoms by means of a five-membered or shorter chain to give a *trans*-arrangement. Therefore, in the cyclo-olefin homologous series we encounter only one isomer up to cycloheptene, the isomer with the *cis*-configuration. Only a chain of at least six carbon atoms is capable of bridging the double bond and joining both carbon atoms from the opposite sides of the double bond plane, thus forming a stable configuration. Consequently, we have two cyclo-octenes, the *cis*- and the *trans*-isomers. If, by some means or other, we bring about an equilibrium isomerization reaction, in the course of which one geometrical isomer is transformed into the other one, we obtain a mixture which in the eight-membered series contains only an insignificant amount of *trans*-isomer. Therefore, in the case of cyclo-octene the *trans*-configuration has a high energy requirement and we may say that such a ring is subjected to great strain (tension). The amount of *trans*-isomer in the equilibrium mixture increases with increasing ring size, and for an eleven-membered cyclo-olefin the equilibrium mixture contains more *trans*-isomer than *cis*-isomer (3). The energy differences of larger rings are determined only by nonbonded interactions, which will be discussed in the chapter on conformational analysis.

Isomerism of Compounds with Cumulative Double Bonds

When a compound contains several double bonds, not immediately adjacent to each other, isomerism is encountered on every suitably substituted double bond quite independently. The number of isomers is given by the number of possible combinations. A different situation arises if the double bonds are directly adjacent; two cumulative double bonds form a rigid system, in which the planes of the two double bonds are at right angles to each other. Thus also the four substituents attached to the ends of the cumulative system are contained in two mutually perpendicular planes; at the intersection of these there are three atoms linked by two cumulative double bonds. The whole system may be illustrated by the formula

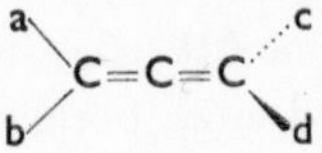

$$\begin{matrix} a \\ & C{=}C{=}C \\ b \end{matrix} \begin{matrix} c \\ \\ d \end{matrix}$$

The carbon atoms and the substituents a and b are coplanar and in this case are in the plane of the paper. The substituents c and d are in the plane perpendicular to the plane of the paper and passing through the carbon atoms, one of which projects above the plane of the paper whereas the second one recedes below it. Each substituent is therefore equidistant from both substituents on the other end of the three-carbon chain, and even if all the substituents are different, this compound has only one geometrical isomer. Such molecules may, however, show optical isomerism.

A third cumulative double bond will again return the substituents on one end of the system into the plane defined by the bonds of the substituents on the other end. The substance then displays two geometrical isomers, *cis*- and *trans*-, like a simple olefin. The same is true for every other system with an odd number of cumulative double bonds since the end substituents are always in one plane without regard for the length of the chain. On the other hand, in systems with an even number of cumulative double bonds the substituents on the ends of the chain are in two planes at right angles to each other; optical isomerism is possible in favourable cases, but *cis-trans* isomerism is not.

Triple Bond

The triple bond is a spatially very simple system. The remaining bonds, attached to the two trebly linked carbon atoms, are on a straight line defined by the centres of these atoms, and point in opposite directions. Four atoms in total are thus kept on one straight line and only at the end of this four-carbon array is the linearity broken. A triple bond may therefore be incorporated into a ring only if the ring has a sufficient span – as is true for a double bond in a *trans*-configuration. The smallest cyclic acetylene prepared hitherto is cyclo-octyne and it is improbable that a stable lower homologue could exist. Large rings may of course incorporate several triple bonds.

Geometrical Isomerism at Multiple Bonds between Non-Carbon Atoms

The double bond between two atoms of nitrogen, or between a carbon and a nitrogen atom, is spatially arranged in the same manner as the double bond between two carbon atoms. The place of the fourth substituent is always occupied by the free electron pair of the nitrogen atom. With regard to the considerable freedom of movement of the electron pair, the stability of geometrical isomers is, in this case, much less than the stability of *cis*- and *trans*-olefins. Nevertheless, two isomers were isolated in the azobenzene series, differing by the distance between the aryl groups. In order to express the reduced stability of these isomers in comparison with olefins, the equivalent prefixes *syn* (for compounds having substituents on the same side of the double bond plane) and *anti* (for the opposite isomer) were used instead of prefixes *cis* and *trans*, *e.g.*:

$C_6H_5-N=N-C_6H_5$ (both C_6H_5 on the same side)

syn-azobenzene or *cis*-azobenzene

$C_6H_5-N=N-C_6H_5$ (C_6H_5 groups on opposite sides)

anti-azobenzene or *trans*-azobenzene

Since one isomer is easily transformed into the other one, we usually encounter an equilibrium mixture of both forms and we have to be very careful if we aim to react truly geometrically uniform substances. Some authors assume the same isomerism for diazotates and isodiazotates. In this case, however, it is possible that we are faced with positional isomerism instead of geometrical isomerism.

Compounds with a double bond between carbon and nitrogen atoms include oximes, hydrazones, semicarbazones and other nitrogenous derivatives of carbonyl compounds. As with azocompounds, the designation *syn* instead of *cis* and *anti* instead of *trans* has become customary for oxime isomers; the reference groups being hydrogen and hydroxyl. (In older literature, the designation is the opposite.)

Thus benzaldehyde oxime, as well as a number of other compounds, has been separated into the *syn*- and *anti*-forms:

$(C_6H_5)(H)C=N-OH$ (OH on the side of C_6H_5)

anti-benzaldehyde oxime

$(C_6H_5)(H)C=N-OH$ (OH on the side of H)

syn-benzaldehyde oxime

Like azocompound isomers, oxime isomers are rather unstable and under the conditions of many reactions, especially in an acidic medium, the less stable form is transformed into the more stable one. The designation *syn* and *anti* is not entirely satisfactory for aldehyde oximes, as can be seen from the discrepancies in the literature. In the case of non-symmetrical ketones, it becomes completely obscure; thus it is always necessary to state in the designation which of the groups are in the position in question, *e.g.* "acetophenone oxime with the phenyl and hydroxyl groups in *syn*-position". This designation is very cumbersome. Oximes of symmetrical ketones cannot form geometrical isomers at all, in a similar manner to olefins having the same substituents on one end of the double bond.

Optical Isomerism

Basic Concept

We may imagine molecules which are formed by the same atoms, in the same number, order and distance, but are not absolutely identical. They differ in the same manner as an asymmetric object and its mirror image (the left and right hand, a left and right helix, *etc.*). Of course, this is true only when the molecule does not manifest reflection symmetry (mirror symmetry), *i.e.* if it is not absolutely identical with its mirror image. Such a model is designated as **dissymmetric**, or recently and more exactly as **chiral**, and the property of a model without mirror symmetry as **chirality**.

Since the spatial configuration of these two isomeric molecules is otherwise the same, these molecules have the same energy content and identical scalar physical properties (*e.g.* the melting point, boiling point, refractive index, density, *etc.*). The chirality of the molecule may influence only the vectorial properties of the molecule, and is therefore connected with the capability of rotating the plane of polarized light. The isomerism observed as a result of the chirality of the molecule is therefore often called **optical isomerism**. Two chiral molecules, differing only as an object and its mirror image, rotate the plane of polarized light by the same angle but in the opposite sense (to the right or left). Pairs of such substances are called **enantiomers** or **antipodes**.

The influence of some substances on the plane of oscillation of polarized light has been known for a long time, and is designated as **optical activity**. The origin of optical activity may vary. Certain inorganic compounds are optically active in the solid state only; after being melted or dissolved their optical activity disappears and is therefore evidently connected with their crystalline structure. In other cases one observes the rotation of the plane of oscillation of polarized light during the passage of the beam through a certain compound in a powerful magnetic field. The angle of rotation is proportional to the intensity of the magnetic field and is thus a transient property. Some organic compounds are, however, optically active in the melt, in the solution and even in the gaseous phase, as had been observed by Biot in 1815. Their optical activity therefore has to be a property of the molecules themselves and must be determined by their chiral structure. Arising from such considerations, Le Bel and van't Hoff formulated a theory assuming that molecules are three-dimensional structures and thus laid the foundations for the concepts concerning their structure, as we have outlined them and as they are generally accepted today.

In order to establish whether a molecule with a given structure is capable of existing in enantiomeric forms, we determine whether it is superimposable with its mirror image. If this is not the case, the substance may exist in two optically active forms as enantiomers. This unequivocal procedure is perceptively exacting; therefore, we attempt to simplify it by taking into account only certain elements of symmetry, the presence of one or more of which gives the molecule symmetry and eliminates the possibility of optical activity of the substance. These are as follows: the plane of symmetry, the centre of symmetry and, generally, the rotation-reflection (alternating) axis of symmetry (see p. 30). Each of these elements alone may eliminate the optical activity of the substance. Nevertheless, a chiral molecule may have an *n*-fold rotational axis of symmetry or be dihedrally symmetric. Only the structures which are non-superimposable with their image, neither by rotation around the axis of symmetry, nor by reflection in the plane of symmetry, nor by a combination of reflection and rotation, are asymmetric. For example, C_{abcd} is an asymmetric carbon atom. Compounds of this type manifest optical activity. The asymmetry of the molecule in the stated sense is thus a sufficient, although not necessary, condition for the existence of enantiomers. The necessary and sufficient condition is the impossibility of transforming the model of the molecule into its identical image by reflection by a plane (Pasteur), *i.e.* chirality.

Asymmetric Carbon Atom

All univalent and bivalent atoms have a plane of symmetry and therefore cannot cause chirality of the molecule. Trivalent and tetravalent atoms with at least two identical substituents also have a plane of symmetry. Only a trivalent or tetravalent atom having all substituents different from each other does not have a plane of symmetry and may – but need not – cause asymmetry of the whole molecule. Such an atom (usually a carbon atom) is called an **asymmetric atom** and is the most frequent cause of chirality of a molecule and thus also of optical activity. (Therefore, we sometimes encounter the inexact term "optically active atom".) In principle, the substituents of an asymmetric atom may be arranged in two ways: "right" and "left", in the same sense as when we speak of a left-hand or right-hand helix. Molecules with one asymmetric atom thus appear as two enantiomers with the same physical properties but with optical activity of the opposite sense. Since the absolute value of rotation is the same in both cases, a mixture of the same amounts of both enantiomers is optically inactive; the activity of one enantiomer is completely cancelled by the activity of the other one. The product is then termed a **racemate** (racemic mixture). Often, however, the enantiomers in equimolar amount do not just form physical mixtures but real new compounds (racemic compounds), which differ in their basic physical properties, such as melting point, solubility and even boiling point, from both enantiomers. The most reliable method of distinguishing a racemic mixture from a racemic compound is to plot a diagram of the dependence of the melting point on the composition of the mixture (by thermal analysis). If the melting point of one enantiomer decreases continuously with the addition of the other one until a minimum is reached at the point of equivalence, and then rises again in the same manner, this indicates a mixture of two substances crystallizing independently in their own crystal lattices; in principle, such a mixture could be resolved by sorting the sufficiently developed crystals by hand. If the melting point remains constant throughout the whole range of the ratios of enantiomers in the mixture, we encounter a relatively rare case of isomorphism; the enantiomers form so-called solid solutions, where both components are included at random in a common crystal lattice. Such a mixture cannot be resolved mechanically. Finally, the most frequent case occurs if the curve shows two minimum values and a marked maximum value at the point of equivalence; the latter indicates the formation of a racemic compound. This type of racemate also cannot be divided by mechanical means into its enantiomers. The three basic types of diagram mentioned are schematically represented in Fig. 2. The methods of obtaining optically active materials from racemates, in other words the resolution of racemates, will be discussed in a special chapter.

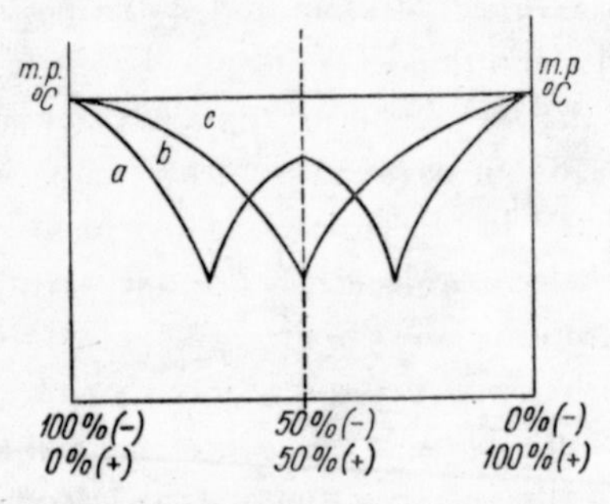

Fig. 2. Basic types of dependence of melting point of mixtures of enantiomers on composition.

a – the racemate forms a compound; b – the racemate is a physical mixture; c – the racemate forms a solid solution, the enantiomers are isomorphous.

In the case where the molecule has only one asymmetric atom, it is chiral without exception. This need not be the case if the number of asymmetric atoms in the molecule is larger. The number of constitutionally identical isomeric individuals increases with each additional asymmetric atom. If a compound with one centre of chirality can have two stereoisomers – if we do not consider the racemate a chemical individual – then the number of isomers may double with each additional element of chirality. A compound with two centres of chirality can thus generally exist as four optically active isomers; a compound with three centres may have eight optically active isomers and finally a compound with *n* centres of chirality will have *2* isomers. Assuming that each centre of chirality is either positive (+) or negative (−), just as the rotation of enantiomers is either positive or negative, we may designate the four isomers of a compound with two asymmetric atoms in the following general way:

$$\begin{matrix} +a & -a & +a & -a \\ +b & -b & -b & +b \end{matrix}$$

Since + is the mirror image of −, the compound designated $\begin{matrix} +a \\ +b \end{matrix}$ is the mirror image of compound $\begin{matrix} -a \\ -b \end{matrix}$, the mirror images of the other two compounds correspond in the same manner. The set of four isomers of a substance with two asymmetric atoms therefore consists of two pairs of enantiomers. Whilst the enantiomers are physically identical, except for the opposite sense of their optical rotations, two components from different pairs of enantiomers may differ substantially. These are not enantiomers but **diastereoisomers** and, as a rule, may therefore be separated by usual physical methods. Like compounds with one asymmetric atom, an equimolar amount of enantiomers with two asymmetric atoms forms a racemate. Thus, the two pairs of enantiomers form in this case two different racemates. In general, the number of racemates must therefore equal one half of the number of possible, optically active substances; for a compound with n asymmetric atoms the number of racemates is given by the expression $2^{(n-1)}$.

The above considerations hold only for compounds having different individual centres of chirality. On the other hand, if some of the centres have the same substituents, the molecule as a whole may, in a certain arrangement, have a plane of symmetry and become optically inactive. If, for example, a compound has two identical asymmetric atoms, two of the four possible combinations become identical. This also follows from the general diagram formulated as in the preceding case, with the difference that centre "b" is replaced by centre "a":

$$\begin{array}{cccc} +\mathrm{a} & -\mathrm{a} & +\mathrm{a} & -\mathrm{a} \\ +\mathrm{a} & -\mathrm{a} & -\mathrm{a} & +\mathrm{a} \end{array}$$

The first two compounds are optically active and are also enantiomers of each other, in agreement with the general case of a compound with two asymmetric atoms of different structure. The second two combinations are, however, in this case identical and represent therefore only one isomer. Since the centres of chirality are identical, we may imagine that the rotation effect bestowed upon the molecule by one centre is just equalized by the contribution of the other part of the molecule, in the opposite sense. Although such internally compensated compounds have centres of chirality, their molecules as a whole nevertheless have a plane of symmetry, and they are called ***meso*-compounds.** Unlike racemates, their zero optical activity is not caused by the mixing or reacting of two different substances (enantiomers), and therefore they cannot be resolved into enantiomers. Because of this they are sometimes called the "non-resolvable form". The first two isomers, which are optically active, form a racemate like any two enantiomers, this being a further physically different form.

The best example of a compound with two identical centres of chirality is tartaric acid COOH . CHOH . CHOH . COOH. The second and third carbon atoms are linked by each of their four valencies to a different atom or group of atoms, and are therefore asymmetric. However, the substitution is absolutely the same for both, with the exception of the spatial arrangement. Their arrangement may be either positive, +, or negative, −. The acid including asymmetric centres with opposite contributions $\begin{pmatrix} + & \text{or} & - \\ - & & + \end{pmatrix}$ is optically inactive and known as mesotartaric acid. The acid with centres of the same configuration $\begin{pmatrix} + & \text{or} & - \\ + & & - \end{pmatrix}$ is optically active and appears as two enantiomers, known as right- and left-handed tartaric acid. The compound formed by equimolar amounts of right-handed and left-handed tartaric acids is optically inactive like mesotartaric acid; however, it may be resolved into enantiomers. The compound is trivially known as racemic acid.

A compound with three asymmetric atoms may in general exist, as we already know, in the form of eight optically active isomers and four racemates. The number of isomers is given by the number of combinations:

+a	−a	+a	−a	+a	−a	+a	−a
+b	−b	+b	−b	−b	+b	−b	+b
+c	−c	−c	+c	+c	−c	−c	+c
1	*2*	*3*	*4*	*5*	*6*	*7*	*8*

However, if two centres are identical (if, for instance, a = c, then the number of optically active isomers is reduced. Namely, in the case of the compounds designated *1*, *2*, *5* and *6* in our diagram, the central atom (b) will thus cease to be a centre of chirality as it is linked to two identical groups. Compound (*1*) will thus become identical with compound (*5*) and compound (*2*) with compound (*6*). Thus, only two enantiomers will remain, instead of four isomers. The central atom of the remaining four compounds continues to be asymmetric but the molecules have a plane of symmetry and are therefore optically inactive. Compound (*3*) is in this case identical with compound (*4*) and compound (*7*) with compound

(*8*). The central atom, although asymmetric, contributes no optical activity to the molecule and is therefore designed a **pseudoasymmetric atom**. A compound with three asymmetric atoms, two of which are identical, may exist therefore as two optically active enantiomers (*1* = *5* and *2* = *6*) with the corresponding racemate and in two optically inactive, non-resolvable configurations. Trihydroxyglutaric acids or pentitols serve as examples of such symmetrically built molecules with three asymmetric atoms.

The number of isomers is similarly reduced in the case of compounds with a larger number of asymmetric atoms, provided some of them are identically substituted and those molecules which have a plane of symmetry will be optically inactive. All that has been stated for compounds where the centres of chirality (asymmetric atoms) are directly linked to each other (tartaric acid) is also valid for compounds where the centres are separated from each other by a carbon or non-carbon chain. For example, the same isomerism as that of tartaric acid (*i.e.* one *meso*-form, two optically active enantiomeric forms and one racemate) is present in the case of the bis-α-phenylethylamide of malonic acid:

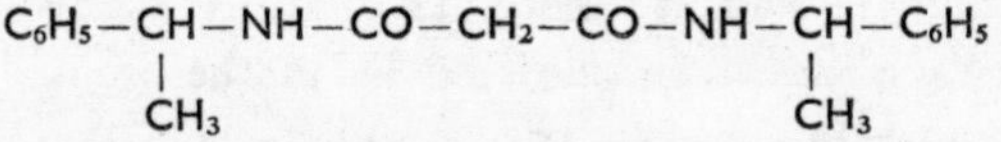

$$C_6H_5-\underset{\displaystyle CH_3}{\underset{|}{CH}}-NH-CO-CH_2-CO-NH-\underset{\displaystyle CH_3}{\underset{|}{CH}}-C_6H_5$$

Such a compound may also exist in optically active forms, provided the configurations of the α-phenylethyl groups are identical, or in the *meso*-form, if they are opposite. However, every small nonsymmetrical change in any *meso*-compound may cause the appearance of optical activity. Consequently, a monoester or any similar derivative of mesotartaric acid may appear in an optically active form or as a racemate which can be resolved into enantiomers. Optical activity in this case again disappears after liberating the acid or forming a symmetrical diester. Similarly, for example, the benzylidene derivative of the previously mentioned *meso*-diamide of malonic acid (*I*) may exist in optically active forms:

$$(C_6H_5)(H)C=C(CONH\underset{\displaystyle CH_3}{\underset{|}{CH}}C_6H_5)(CONH\underset{\displaystyle CH_3}{\underset{|}{CH}}C_6H_5)$$

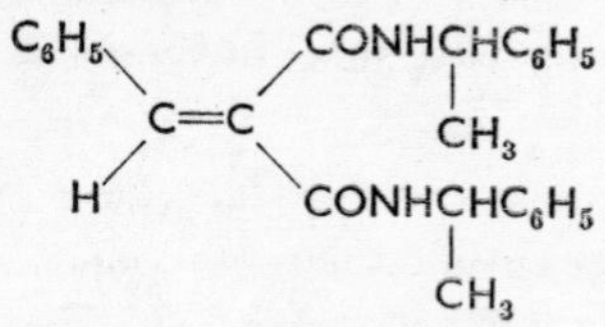

I

The phenyl group is in a *cis*-position to one substituent and in a *trans*-position to the other.

The above conclusions are valid not only for acyclic but also for cyclic compounds. Vicinally disubstituted cycloalkanes with an arbitrary ring span with identical substituents are analogous to the tartaric acids. The *cis*-derivative constitutes the *meso*-form, optically inactive and non-resolvable; the *trans*-isomer may exist in the right-handed and left-handed optically active forms and as a racemate. The optical activity of the compound or the resolvability of the racemate are in this case an unequivocal proof of the *trans*-configuration of the substituents in the molecule.

1,3-Disubstituted cyclobutanes have a plane of symmetry passing through the atoms in question and are therefore optically inactive in the *cis*- and *trans*-forms. In 1,3-disubstituted compounds with a five-membered ring, the situation is analogous to compounds with vicinal substituents.

Both *cis*- and *trans*-forms of 1,4-disubstituted cyclohexanes also have a plane of symmetry and are therefore optically inactive. With larger rings, isomerism is again encountered as in the vicinally substituted cyclic compounds. In general, trisubstituted cycloalkanes have three asymmetric carbon atoms and the respective number of isomers. Considering the combination of different positions, different ring sizes and different substituents, it becomes clear that the number of isomers is so large as to make individual analysis impossible in this book. However, trisubstituted cyclopropane is especially interesting. If two substituents are identical, the third carbon atom may be pseudoasymmetric, and the compound is thus encountered – as is trihydroxyglutaric acid – in two *meso*-forms and two optically active enantiomeric forms which may form a racemate together. If all three substituents are the same, the compound has two isomers, one with all the substituents located on one side of the ring, and the other one with two substituents on one side and one on the other side. Both isomers have a plane of symmetry and are thus optically inactive.

Heterocyclic compounds often do not have the plane of symmetry appearing in analogous carbocyclic compounds. Whereas monosubstituted alicyclic compounds have a plane of symmetry and are therefore optically inactive, monosubstituted derivatives of pyrrolidine, tetrahydrofuran, 2- or 3-substituted piperidine and tetrahydropyran derivatives as well as analogous heterocyclic compounds may exist as two enantiomers and as a racemate. Only when six-membered heterocycles are substituted in position 4 are they symmetrical and optically inactive. Similarly disubstituted derivatives, even if they have identical substituents, do not as a rule appear in *meso*-forms like *cis*-disubstituted cycloalkanes, but in two pairs of optically active forms, with the corresponding two racemates. The substituents are symmetrically grouped with respect to the heteroatom only if the *cis*-isomer is an optically inactive *meso*-form (*e.g.* 2,5- or 3,4-disubstituted pyrrolidines and tetrahydrofurans or 2,6- and 3,5-disubstituted piperidines and tetrahydropyrans).

Non-Carbon Asymmetric Atoms

Until now, our considerations have been of a general nature, but the examples shown gave the impression that a carbon atom has to be situated at the centre of chirality. However, a number of compounds are known where the asymmetric atom is an atom of nitrogen, phosphorus, arsenic, sulphur, silicon or boron, not considering the metal complex compounds with a considerably different structure.

A neutral nitrogen atom is trivalent; its three valencies are, however, not coplanar, but form the edges of a low pyramid, substantially similar to the configuration adopted by three valencies of a carbon atom. The position of the fourth carbon valence is, in the nitrogen atom, occupied by the free electron pair. In principle, compounds in which three different substituents are attached to the nitrogen atom could thus exist in optically active forms. However, since the configurations of the nitrogen atom are not sufficiently stable, as a result of the presence of the relatively mobile electron pair, the isolation of such a compound has not been successful, provided the substituents have not been held in their positions by the rigid backbone of the whole molecule. Troeger's base *II* (produced by reacting *p*-toluidine with formaldehyde) is optically active because inversion of nitrogen is structurally prohibited. The preparation of an optically active form of the indolocarbazole derivative *III* has not been successful.

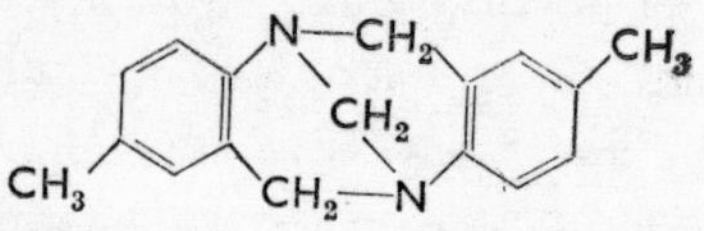

II 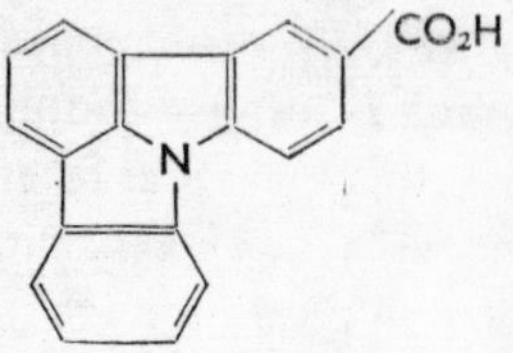 III

Whilst tertiary amine bases are resolvable into optical enantiomers in very rare cases only, optically active quaternary ammonium salts are quite common and stable. The nitrogen atom in the quaternary salt is practically identical with the carbon atom from the aspect of bond distribution and almost as regards its dimensions; therefore, the same principles hold for its asymmetry. The first optically active ammonium salt, methylallylphenylbenzylammonium iodide, was prepared at the end of the last century and since then many similar salts have been prepared, as well as compounds with a larger number of asymmetric nitrogen atoms, like the bis-methiodide of diethyldiphenylethylenedia-

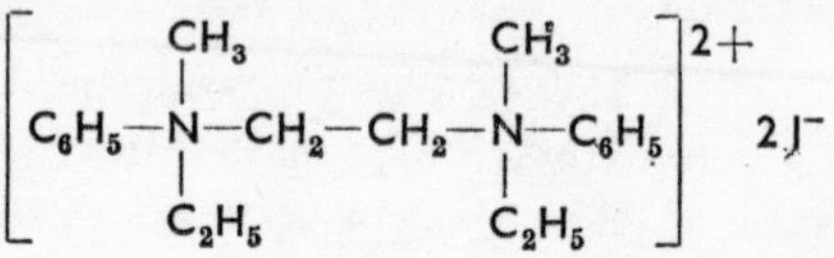

$$\left[C_6H_5-\overset{\displaystyle CH_3}{\underset{\displaystyle C_2H_5}{N}}-CH_2-CH_2-\overset{\displaystyle CH_3}{\underset{\displaystyle C_2H_5}{N}}-C_6H_5 \right]^{2+} 2J^-$$

IV

mine (*IV*). The compound contains two asymmetric atoms identically substituted, and therefore shows the same type of isomerism as the tartaric acids; it exists as a *meso*-form, as two optically active enantiomeric forms and as a racemate.

The fourth nitrogen valency in trialkylamines may also be occupied by an oxygen atom. If the alkyl groups differ, the nitrogen atom of the trialkylamine oxide is also asymmetric. Therefore, completely stable, optically active compounds of the methylethylpropylamine oxide type have been prepared.

In contrast to tertiary ammonium bases, trisubstituted phosphines are extremely configurationally stable. Optically active methylpropylphenylphosphine even withstands vacuum distillation. It loses its optical activity (racemizes) only when distilled under normal pressure. Quaternary phosphonium salts may be obtained in optically active forms, similarly to ammonium salts, which is not surprising. Asymmetrically sub-

$$\overset{(+)}{(CH_3)_3N}-C_6H_4-\underset{\underset{O}{\|}}{\overset{\overset{CH_3}{|}}{P}}-OCH_3 \qquad\qquad C_2H_5-\underset{\underset{OH}{|}}{\overset{\overset{OC_2H_5}{|}}{P}}=S$$

V *VI*

stituted phosphine oxides may also be optically active. Among phosphorus compounds, however, further types of asymmetric molecules are also encountered, *e.g.* the derivative of phosphinic acid *(V)* or of thiophosphonic acid (*VI*).

As an example of an optically active arsenic compound, a benzylethylpentylphenylarsonium salt has been prepared.

The free electron pair in compounds containing trivalent sulphur (as in the case of phosphines) does not have such mobility as in the compounds of nitrogen and, therefore, optically active sulphonium salts, sulphoxides, esters of sulphinic acids, and sulphinylamines have been successfully prepared. The stability of configuration of these trivalent asymmetric atoms, compared with nitrogen atoms, is probably caused by the fact that the pyramid formed by the sulphur or phosphorus valencies is somewhat taller than that of the nitrogen atom, because of the larger inter-atomic distances. Consequently, interconversion into the second configuration, *i.e.* transition by way of the symmetrical planar configuration, is less facile.

Symmetry Elements Eliminating Optical Isomerism

We have already seen that even compounds having asymmetric atoms with a stable configuration sometimes cannot be obtained in optically active form. This happens when the molecule has a **plane of symmetry** and such compounds are known as *meso*-forms. However, in some cases the compound has asymmetric atoms, does not have a plane of symmetry and nevertheless is optically inactive. The example most frequently cited is *trans*-3,6-dimethyl-2,5-piperazinedione (*VII*); compounds of this type have a **centre of symmetry**, *i.e.* a point midway between all pairs of analogous atoms. If a molecule has a centre of

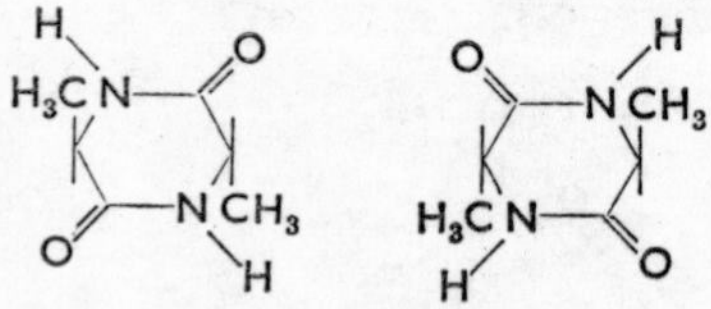

VII

symmetry it cannot be optically active. We can prove this by constructing a model of the molecule and its mirror image; by rotation of the mirror image through 180° we obtain a model absolutely identical with the initial one. In this special case we may also imagine that *trans*-3,6-dimethyl-2,5-piperazinedione was formed by condensation of (+)-alanine and (−)-alanine, so that the rotations of both components cancel one another, as in *meso*-compounds. The conclusion that the presence of a centre of symmetry prevents optical activity is, however, generally valid.

Yet another possibility for the cancellation of molecular asymmetry had been predicted quite a long time ago for compounds which have individual asymmetric atoms and have neither a plane nor a centre of symmetry but which incorporate a **fourfold rotation-reflection axis of symmetry**. A rotation-reflection axis (also alternating axis) of symmetry is a straight line around which an object may be rotated in such a manner as to obtain its mirror image (in contrast to the usual axis of symmetry, where by rotation we obtain an image identical with the initial object). By the order of the axis we understand the number of positions of the object in the required position encountered during one 360° turn. Generally, in the case of an *n*-fold axis one obtains the original arrangement *n* times in the course of one turn; the smallest angle of rotation is thus 360°/*n*. With a fourfold axis of symmetry we thus have to rotate the molecule by 90° to obtain the identical arrangement, or, in the case of an alternating axis, the mirror image. Only recently, however, has a compound with a fourfold rotation-reflection axis as the only element of symmetry been prepared, and proof provided of its optical inactivity. It is a *meso-trans, trans*-3,4,3′,4′-tetramethyldipyrrolidinium salt.

Of course, there are compounds which have no asymmetric atom whatsoever and nevertheless are totally asymmetric and thus capable of rotating the plane of polarized light. In the same sense as we have spoken of the centre of chirality (situated at the centre of the asymmetric atom) in the instance of compounds with an asymmetric atom, we may, in the case of certain compounds, speak of the **axis of chirality or plane of chirality**, even if the geometrical meaning of these terms is difficult to define. They are usually defined as a straight line or plane which would become the axis or plane of symmetry if some of the different substituents were made the same.

Whilst for models with a centre of chirality all the four groups have to be different, for ones with an axis of chirality a smaller number of differences is sufficient, as we will see later on; for models with a plane of chirality, even one difference is enough.

A special example of chirality is **helicity**, *i.e.* the arrangement of atoms or groups on an imaginary helix. Helicity is defined by its sense, which we consider as right-handed if a point recedes from the observer when moving clockwise along the helix, and left-handed when it recedes from the observer during anticlockwise motion. The helicity sense corresponds to the generally accepted sense of the left-handed and right-handed pitch of a screw. It does not change if we select the opposite end for observation.

During the analysis of the spatial structure of molecules we often encounter the term **symmetry**. For clarification, we will now give a survey of its main types—**point groups**.

The classification is based on three elements of symmetry: the plane of symmetry, the *n*-fold axis of symmetry and the *n*-fold rotation-reflection axis of symmetry. The centre of symmetry does not have to be considered independently as it is geometrically identical with the twofold rotation-reflection axis of symmetry. On the basis of the above three elements of symmetry, we may classify molecular models into two main groups and five categories. For each category, we will introduce the types of symmetry belonging to it with the help of the Schoenflies symbols, as follows:

A. Categories without mirror symmetry (chiral)

1. no element of symmetry (C_1)
2. one or more axes of symmetry (C_n, D_n)

B. Categories with mirror symmetry (achiral)

1. a plane of symmetry but no axis of symmetry (C_s)
2. no plane of symmetry, only an n-fold rotation-reflection axis (S_n)
3. a plane of symmetry as well as axes of symmetry (C_{nv}, C_{nh}, D_{nd}, D_{nh}, T_d, O_h)

The symbols in brackets are used for the designation of the individual types of symmetry and have the following meanings:

C_1 – group of asymmetric models (onefold axis of symmetry not considered because every object has this axis).

C_n – group of models having only n-fold axis of symmetry ($n \neq 1$).

D_n – besides the main n-fold axis of symmetry, the model has n twofold axes of symmetry in the plane perpendicular to the main axis, which of course must not be the plane of symmetry ($D_n = C_n + nC_2$). This type is commonly known as dihedral symmetry.

C_s – group of models having only a plane of symmetry (σ).

Examples: chloroethylene, bromocyclopropane.

S_n – rare type without plane of symmetry, with only n-fold rotation-reflection axis of symmetry. Examples given already.

C_{nv} – one n-fold axis of symmetry and n planes of symmetry intersecting in axis of symmetry (v means vertical). The respective planes are designated σ_v($C_{nv} = C_n + n\sigma_v$).

Examples: water (C_{2v}), ammonia (C_{3v}).

C_{nh} – one n-fold axis of symmetry and one plane of symmetry perpendicular to this axis (h means horizontal, $C_{nh} = C_n + \sigma_h$).

Example: *trans*-dichloroethylene.

D_{nd} – the group of dihedral symmetry supplemented with n σ_v planes of symmetry; the planes of symmetry halve the angles of the dihedral C_2 axes and intersect in the main C_n axis.

Example: cyclohexane in chair conformation (D_{3d}).

D_{nh} – dihedral symmetry with n σ_v planes and one σ_h planes.

Example: benzene (D_{6h}).

T_d – tetrahedral symmetry: 4 C_3 axes, 3 C_2 axes and 6 σ planes.

Example: methane.

O_h – octahedral symmetry (cubic): 3 C_4 axes, 4 C_3 axes, 6 C_2 axes and 9σ planes.

Example: cubane.

Axis of Chirality

An **axis of chirality** is encountered in allene derivatives, alkylidene cycloalkanes, spiranes, adamantane and in biaryl derivatives. Whereas the first four groups are characterized by stable constitutional isomerism of the enantiomers, the last group presents a somewhat different phenomenon, usually known as **atropisomerism**.

Optical Isomerism of Allene Derivatives

The spatial arrangement of the cumulative double bonds of allene has already been discussed and we will now repeat only the most important finding, *i.e.* that the four substituents of the allene grouping are situated at the apexes of an imaginary tetrahedron which, in contrast to the tetrahedron formed by the substituents around one carbon atom, is not regular (see Fig. 3). The irregularity of the tetrahedron causes the impossibility of cyclic interchange of the substituents. The distance between pairs of them differs and therefore it is not necessary for all of the substituents to differ, in order to produce asymmetry; it is sufficient to have each substituent different from its nearest neighbour. Group A (in Fig. 3) may thus be identical with group C and group B may be identical with group D; despite this, the mirror image of the molecule is not identical with the object. The molecule is optically active even in this case. One of the first compounds of this type to be actually prepared was 1,3-diphenyl-1,3-di-α-naphthylallene.

The substituents of rings have the same geometry as the double bond substituents. Thus, methylenecycloalkanes or spirocyclic compounds are, when suitably substituted, also optically active. Typical representatives of this group of compounds are 4-methylcyclohexylideneacetic acid (*VIII*) or the dicarboxylic acid (*IX*) derived from spiro[3,3]heptane:

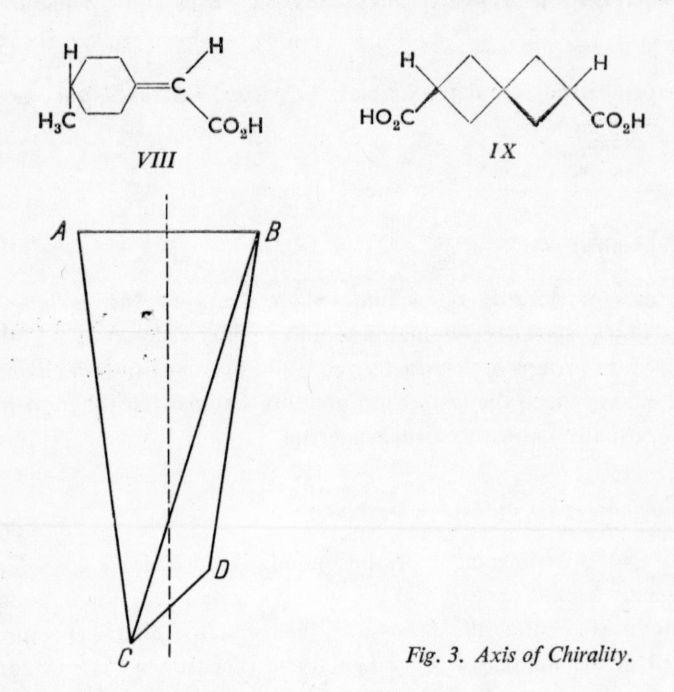

Fig. 3. Axis of Chirality.

Atropisomerism

Atoms and groups of atoms are, as we already know, in principle free to rotate around single bonds. For almost any molecule, we could thus imagine a conformation in which all elements of symmetry would be absent and all substances could thus be optically active. However, the preceding analysis makes it clear that substances may be optically active, only if the molecules may not rotate to achieve a position in which they would have mirror symmetry. Evidently, the reason lies in the fact that free rotation allows a more or less easy transition from one form into the other and the occurrence of even one symmetric for monly creates such a statistical distribution of the individual forms that possible chirality is perfectly cancelled. Therefore, the classification refers to a symmetrical conformation of the molecule, even if it is not the only one and, as a rule, not even the most populated one.

However, it may happen that the bulk of the substituents will prevent the rotation of the molecule in a certain position or will prevent the existence of a symmetrical form of the molecule, although, according to the structural formula, such a form ought to exist. Such a substance may appear in optically active form. This phenomenon is known as **atropisomerism**. The best known case of atropisomerism is the isomerism

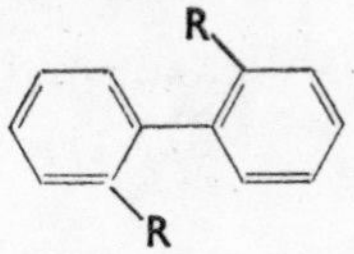

R

R

X

of the *ortho*-substituted derivatives of biphenyl (*X*). If the substituents are sufficiently large, they prevent the rotation of one benzene ring with respect to the other one around the axis passing through the single bond between the aromatic groups. In particular, they make it impossible for the molecule to achieve a symmetrical structure, shown for instance by the above formula. These compounds thus appear in the form of two enantiomers and one racemate. In their geometry, atropisomers approach allene derivatives and, in connection with them, we may speak of the axis of chirality but not of the centre of chirality. Optical activity thus appears even if the substituents are the same, as in the case of the allenes. This means that each aromatic ring has to be asymmetrically substituted with regard to the axis passing through the single bond between the phenyl groups and the *para*-positions.

Whereas in the case of substances with an asymmetric atom, or in the case of allenes, one enantiomer may interconvert into the other one only if a chemical bond is ruptured and formed again, *i.e.* by means of a chemical reaction, in the case of atropisomers the other enantiomer may be formed if the substituent in one *ortho*-position just succeeds in "pushing through" past the smaller *ortho*-substituent on the second ring. The stability of the optical activity of atropisomers may thus serve as a measure of the size (effective volume) of the substituents. Although the hydrogen atom is quite small, optically active compounds exist with two or even three *ortho*-positions of the biphenyl occupied by hydrogen. In the second case the benzene ring that is unsubstituted in the *ortho*-positions has to have a substituent in a *meta*-position. The *meta*-substituent has no influence on the possibility of rotation but it brings about the necessary chirality of the molecule; 3-bromobiphenyl-2′-trimethylarsonium iodide (*XI*) may serve as an example of such compounds.

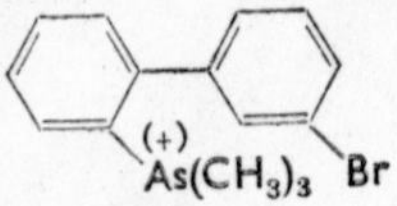

(+)
$As(CH_3)_3$ Br

XI

Usually, however, two, and more often three or four *ortho*-positions are occupied. Measurements have shown that the sum of the effective radii of two groups in *ortho*-positions in different rings has to be larger than 2.9Å in order to prevent the molecule from occupying a symmetrical conformation with the two rings in one plane. The effective radii of certain atoms and groups are shown in Table II. If for instance there is a hydrogen in one *ortho*-position, at least a bromine atom has to be attached to the second ring to stand a chance of preparing an optically active substance. Atropisomerism, however, is not limited to biphenyl

Table II

EFFECTIVE RADII OF ATOMS AND GROUPS (Å)

Group	F	H	OH	CO_2H	NH_2	CH_3	Cl	NO_2	Br	I
Radius	0.39	0.94	1.45	1.56	1.56	1.73	1.89	1.92	2.11	2.20

derivatives; it is encountered in the series of binaphthyls, bipyridyls, phenylpyrrole, bipyrrole, *etc.* Optical activity may also be manifested by compounds in which both aromatic rings are linked through their *ortho*-positions by means of a chain in such a manner as to form a ring larger than six-membered. If a five-membered ring (of the fluorene type) or a six-membered ring (for example the dilactam of

2,2'-diaminobiphenyl-6,6'dicarboxylic acid) is formed, optical activity is eliminated because the small ring enables both aromatic rings to occupy a coplanar position. Compounds with a seven-membered ring may, however, be optically active.

Atropisomers may also be formed by compounds in which one benzene ring is replaced by a substituted ethylene group or another suitable group. For example, an optically active derivative of styrene (*XII*) and substituted anilides of type *XIII* have been prepared:

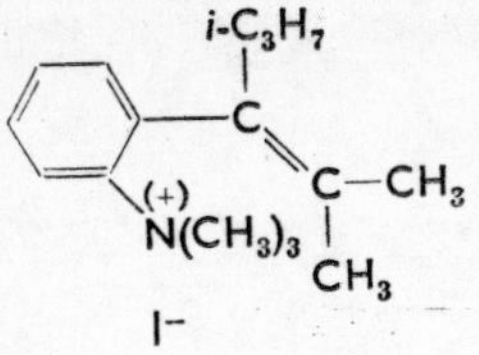

XII

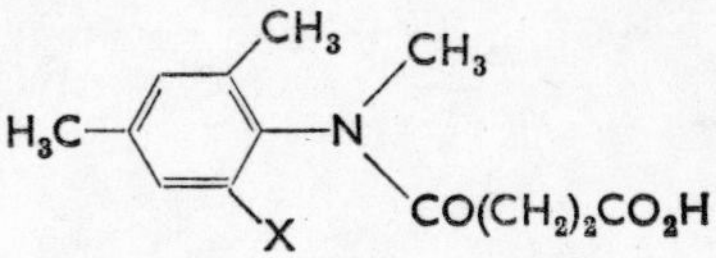

XIII; X=I, Br, Cl, OCH_3, NO_2
(The substituents are arranged in decreasing order of optical stability of the substances)

Plane of Chirality

If a compound is composed of three aromatic rings, substituted as required and linked as in the case of terphenyl, the number of stereo-isomeric forms is multiplied in the same manner as in the instance of compounds with two asymmetric atoms. In general, an asymmetrically substituted terphenyl (*XIV*) may thus exist in four optically active forms, *i.e.* in two diastereoisomeric pairs corresponding to the *erythro*- and *threo*-forms of aliphatic stereoisomers (symmetry C_1). If the central ring has two pairs of identical substituents attached symmetrically (A = B, C = D), one diastereoisomer constitutes an optically inactive *meso*-form (simmetry C_s) whereas the other one comprises two optically active enantiomeric isomers and forms a racemate (symmetry C_2). If the substituents on the central ring are all the same (A = B = C = D),

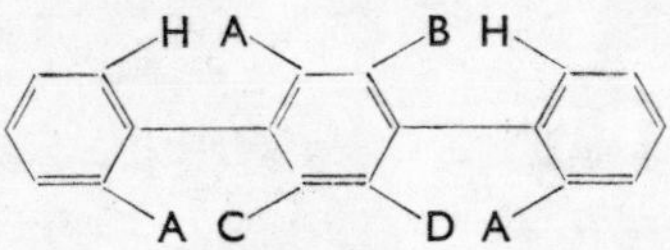

XIV

or pairs of them are the same, symmetrically to the connecting line of the bonds with the other phenyl rings (A = C, B = D), the molecule is achiral and exists in two geometric, optically inactive forms, *cis* (symmetry C_{2v} or C_s respectively), and *trans* (symmetry C_{2h}) only.

The derivatives of hydroquinone polymethylene ethers (*XV*) and substituted paracyclophanes (*XVI*) are optically active, having a **plane of chirality**.

The connecting ring, though large, does not permit the complete rotation of the aromatic rings and the formation of a symmetrical (planar) configuration. Therefore, these compounds exist as two enantiomers and, of course, a racemate.

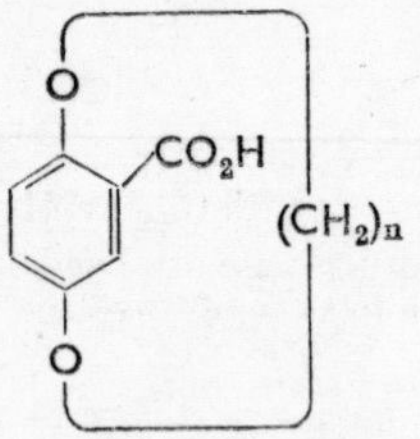

XV 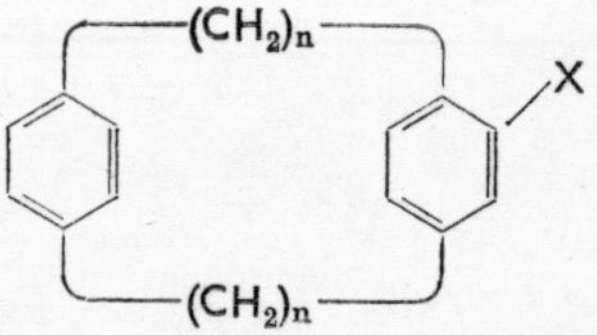 *XVI*

A special case of chirality has been observed in the case of *trans*-cycloöctene (*XVII*) and cycloöctadiene (*XVIII*) (4). In these cases asymmetry is also caused by the fact that the small span of the ring prevents the molecule from achieving a symmetrical, planar configuration, *i.e.* the position shown in formulae *XVII* and *XVIII*. The aliphatic chain protrudes either in front of or behind the plane formed by the atoms linked directly to the ethylene group; it is probable that even the substituents at the double bond are in this case somewhat displaced from the plane and therefore add to the chirality of the molecule.

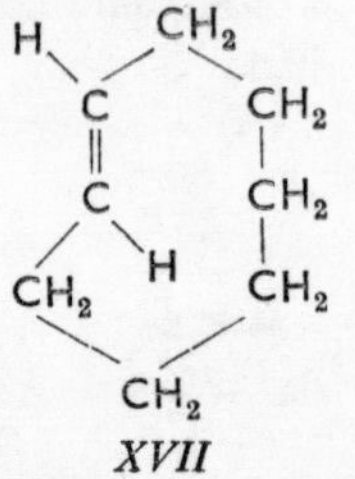

XVII 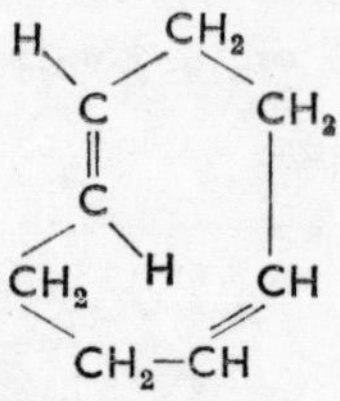 *XVIII*

Finally, aromatic compounds have been encountered, in which suitable substitution forces the molecule to buckle from the most favourable, exactly planar arrangement. For example phenanthrene, which has a planar molecule, as is usual for all aromatic systems, when substituted in positions 4 and 5 by, for instance, methyl groups, becomes somewhat skewed and thus forms two enantiomers. Consequently it was possible to prepare the optically active 4,5-dimethylphenanthrylacetic acid (*XIX*).

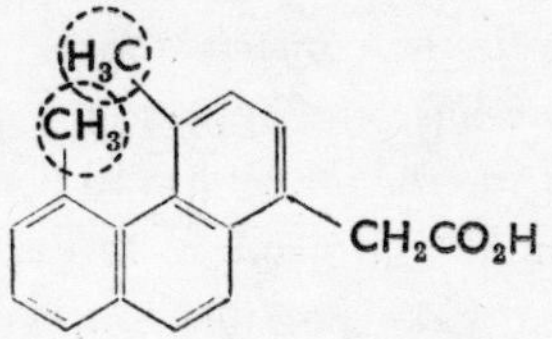

XIX

The chirality of benzophenanthrene derivatives is still more marked, for example in the case of hexahelicene (*XX*), which is characterized by an unusually high optical activity.

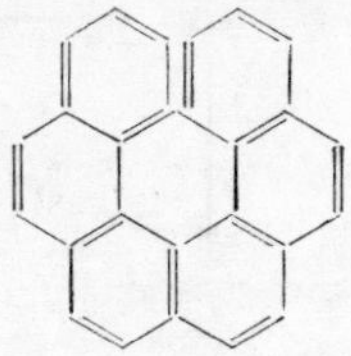

XX

Chemical Formulas and Molecular Models

Until now we have discussed the spatial structure of molecules in general terms. At this stage we have to clarify the relation between the stated general findings concerning the structure of molecules and between real substances. Furthermore, we will have to use the whole concept verbally as well as in writing. First of all, let us therefore distinguish certain terms which could otherwise be interchanged.

The first thing a chemist encounters is a substance, a **chemical individual.** A chemical individual is unequivocally characterized by its physical properties, melting point, boiling point, density, refractive index, by its spectra in various ranges of wavelengths, and finally by its optical activity, which sometimes is the only criterion capable of distinguishing between two otherwise identical individuals.

After isolating and characterizing the chemical individual, the chemist begins to determine its **structure**. The structure may then be expressed by use of a molecular model illustrating the number, position, interconnection, distance, and situation of the individual atoms forming the molecule. However, the construction of actual molecular models is not normal routine because the appearance of the model is difficult to communicate. Therefore, the model has to be transferred on to the paper by means of some suitable projection. Moreover, each substance has to be named in order to make possible communication and recording of findings. At the same time it is necessary to distinguish between stereoisomeric compounds of the same systematic designation with the help of various symbols. The procedure may be summarized by the following diagram:

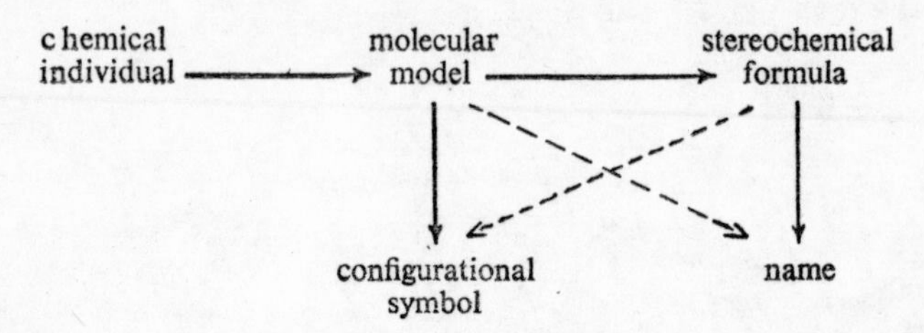

The forming of a relation between the chemical individual and the respective molecular model is designated as the **determination of structure**; in the field of stereochemistry we call this type of activity **correlation** or **determination of absolute and relative configuration.** The relation always has to be created by means of chemical or physico-chemical experiments. On the other hand, the relations between the other terms are based on conventions more or less commonly agreed upon. The name of the compound may be derived, as is shown in the diagram, either from the stereochemical formula or directly from the model. The former method is more usual. Similarly, the configurational symbol also may be derived either from the formula or directly from the model. Because of the fact that the creation of an exact stereochemical formula from the model requires the knowledge of certain conventions, it is more simple to derive the symbol directly from the model. This method is currently gaining importance, especially in general cases of diastereoisomerism. It is necessary to stress that the name and symbol are derived from the molecular model or formula and not from the properties or method of preparation of the chemical individual.

Molecular Models

When it became generally accepted that chemical formulas should express the actual distribution of atoms in the molecule as closely as possible, and not only the numerical ratios in which the individual elements are present or the probable reactivity of substances, an effort was made to characterize the structure with greater precision than could be provided by the formula. Thus, molecular models appeared, and are today considered to be the most illustrative aid for expressing the structure of a molecule. Therefore, in special cases we actually construct an enlarged molecular model. To this end, several sets of more or less precise models are available and each of them has its advantages as well as disadvantages (5).

In principle, we require the dimensions and design of a molecular model set to express as exactly as possible, and to a defined scale, the interatomic distances, the bond angles, and if possible also the occupation of space. In addition the model should illustrate free rotation around single bonds as well as the rigidity of multiple bonds. The models are unable to comply with all the requirements at the same time. They have to be sufficiently rigid and simultaneously they have to permit the small deviations from mean values which we may encounter in reality.

Consequently, two basic types of molecular models, which supplement each other, have been developed: framework type models and "space-filling" type models.

The first, framework type emphasizes mainly the lengths and directions of bonds and neglects the bulk of the atoms. Atoms are represented as point particles only, by means of a small sphere, tetrahedron or other suitable shape, and connected with the help of relatively long wires, springs or plastic rods. A multiple bond may be represented either by a special type of atomic model, or by a multiple connection of normal type models, provided the connecting material allows sufficient bending of the "valences". The advantage of framework type models is in their clarity and especially in the possibility of direct measurement of interatomic, in particular nonbonded, distances, valence angles and torsional angles (see chapter on conformation). Of course, a *conditio sine qua non* is the considerable precision of the model set, especially as regards the exact directional attachment of bonds in the individual atomic models. The best in this respect are the tetrahedron types, permitting the required precision to be maintained during production, or models with precisely adjustable torsion angles, respectively. Wire models have a disadvantage in that they give a false impression of the space requirements of atoms. Therefore, they also do not adequately interpret steric hindrance (effective volumes).

The second type of models is designed to give an enlarged representation of the whole space filled by the atom together with its electron shell, and not only of the nucleus or ideal centre of the atom. The individual pieces are connected by rods and a suitable mechanical connecting device. If the pieces are made of sufficiently elastic material, connecting rods with thicker ends in the shape of small dumb-bells are satisfactory. By joining the sections we obtain a very compact but not very lucid arrangement, on which we may investigate the filling of space with mass and the steric hindrances of groups in individual positions. However, the exact directions of valences and interatomic distance are difficult to ascertain.

The shapes of the original space-filling (Stuart) models were designed on the basis of measurements of the interatomic distances of saturated hydrocarbons. Later on, their dimensions were adjusted by Briegleb for unsaturated and aromatic compounds, in order to improve the representation of the large space which is occupied by free electron pairs. In production, models of this type are very sophisticated as it is necessary to model many different pieces even for the same element with different substituents. For example, models of a carbon atom with four single bonds differ substantially from models of a carbon atom with a double bond, a carbon atom with a triple bond or from an aromatic carbon atom, respectively. A carbon atom in a saturated five-membered ring also differs from the same in an unsaturated five-membered ring, *etc.* If an atomic model of the required type is not available, it is, as a rule, impossible to replace it by another piece, and the model of the molecule cannot be constructed. As in the case of the first type, even an insignificant change of the shape of the constituent parts disfigures the resulting molecule beyond recognition, and it is therefore necessary to manufacture the pieces with high precision, which is very difficult because of their irregular shape. Consequently, only a few manufacturers supply sets which are up to standard.

Projection

Disregarding spatial isomerism, we may represent molecular models very simply by means of the usual structural formulas. In these cases, since the interatomic distances and bond angles are not too important, the usual structural formulas are simplified as much as possible, common groups are represented with the help of generally accepted symbols, and the structure of simple groups remains normally unspecified. The direction of valence lines in usual formulas has no spatial meaning, and formulas may be written in a more or less arbitrary way, provided the number and order of atoms is maintained.

The situation changes in the case of stereochemical considerations, where the direction, distance, and spatial distribution of atoms in the molecule are at stake. For such purposes, a much more exact representation of the model is needed. Therefore, it would seem expedient to have a photograph of the model taken; in some especially difficult and important cases this method is actually resorted to and a photograph of the molecular model or part of it then appears in a scientific paper. However, a photograph tends to be unclear as regards details, gives a one-sided picture and besides that still remains technically pretentious and therefore cannot be applied on a large scale. Somewhat less exact is a perspective drawing of the model. It can be executed in such a manner as to provide the greatest possible clarity of detail; however, the work requires great imagination and skill. The descriptive projection of the model into one, two or three planes is similarly difficult. The drawing is therefore usually done according to certain conventions permitting the simplification of the picture without being detrimental to the precision of the description. First of all, we draw in detail or in exact projection or perspective only those parts of the molecule which are stereochemically interesting, *e.g.* for example the immediate vicinity of a double bond or asymmetric atom, while the other parts of the molecule are represented by collective symbols only. The means of detailed illustration of stereochemically important details vary. The substituents around an asymmetric atom may easily be distinguished by drawing the lines representing the bonds as heavy, light or dotted lines. It is understood that the bonds represented by lines of normal print thickness are contained in the plane of the paper; heavy lines protrude in front of the plane and the dotted ones recede behind the plane. It is immaterial which bonds are chosen as the directions determining the plane, so that the same model may be represented by several absolutely equivalent drawings (see Diagram A 1). The thickening of a line is sometimes emphasized by means of a wedge broadening in the direction of the observer. Formulas treated in this manner are very suggestive and unequivocal without further explanation. The possibility of varying the representations of the same model may be of advantage in that it permits the choice of such an angle of observation which enables us to best grasp the structure of a compound. The disadvantage of the variability is in the distortion of the similarity of related compounds.

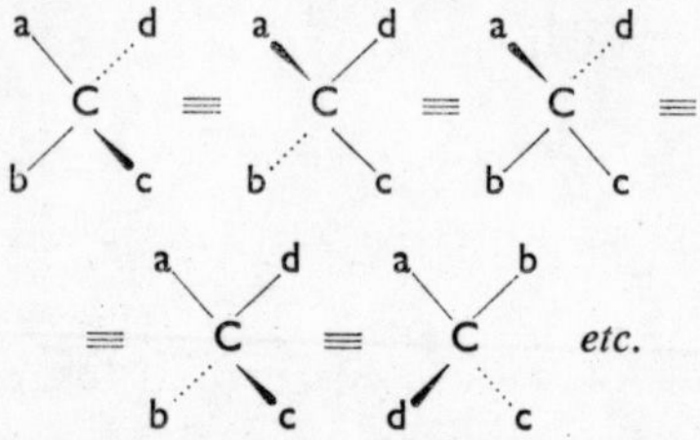

Diagram A 1

The same method may also be used for compounds with several centres of chirality. However, if we are to express the mutual relationship of two asymmetric atoms, projection in the direction of the bond connecting both atoms is very convenient. In this view both asymmetric atoms merge into one point, from which six valences in all project. In order to distinguish between the nearer atom and the farther one, we draw the bonds issuing from the nearer atom from the centre, whereas the bonds of the farther one are drawn from the periphery of a concentric circle only (see Diagram A 2). This type of representation is called the **Newman**

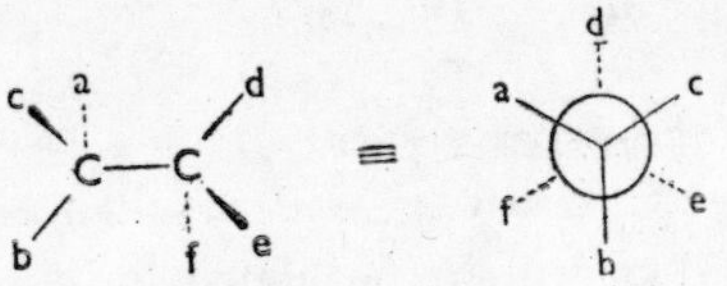

Diagram A 2

projection, and may be used even if both asymmetric atoms are not directly attached to each other. In such a case we imagine that all the atoms connecting the asymmetric atoms lie on the axis of observation and thus merge into one point. The bonds issuing from the former are neglected as sterically uninteresting. Such a projection has to be accompanied by a normal structural formula or by explanatory text.

A commonly used method is the **Fischer projection,** which is used mainly for compounds with several asymmetric atoms, especially in carbohydrate chemistry, but may be applied quite generally. In contrast to the preceding methods, the Fischer projection has very strict conventions and allows very few, and sometimes only one, means of representation of a molecule. Each asymmetric atom is depicted in such a manner as to project two of its bonds on to the plane of the paper in a vertical line and the other two in a horizontal line. The bonds with a vertical projection have to recede from the observer whereas the horizontal ones have to project towards him. The asymmetric atom itself is contained in the plane of the paper and is usually not written out in the projection. Its position is determined by the point of intersection of the horizontal and vertical lines. Conventionally, we arrange the principal carbon chain of the molecule in the vertical line. In the same manner we can represent a number of mutually attached or isolated asymmetric atoms. As a rule for the projection, we consider each asymmetric atom separately. For example, the model appertaining to (+)-tartaric acid is represented according to Diagram A 3 (which shows also the perspective view).

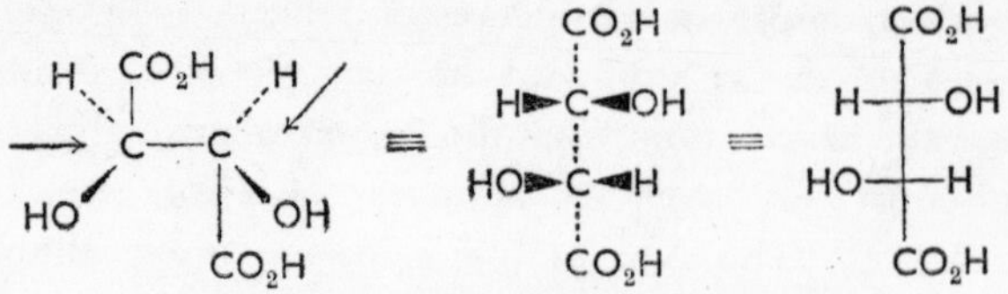

Diagram A 3

The bond between the two asymmetric atoms has to be understood as receding from atom $C_{(2)}$ to $C_{(3)}$, and simultaneously from $C_{(3)}$ back to $C_{(2)}$. Understanding will be facilitated by imagining the chain of carbon atoms forming a broken arc turned so that the ends recede from the observer, as well as by the fact that each asymmetric atom is arranged individually for scrutiny. Consequently, we do not usually observe all the asymmetric atoms at the same time but each of them separately.

The Fischer projection provides very lucid, and at the same time absolutely unequivocal, representations of a molecular model. Its disadvantage is the considerable distortion of the actually most probable conformation of the molecule, as is, however, true for all projections.

The **Haworth projection** is very convenient for the representation of simple cyclic molecules. A closed chain of atoms, a ring, is represented according to Haworth always in a planar configuration and projected on to the paper set at a certain angle. With this method of projection, one part of the ring is nearer to the observer, the other one is farther away. The side nearer in the projection is usually emphasized by thickening the respective valence lines. If the nearer side is not thickened, we take the bottom part of the projection as the nearer one. The substituents are situated on perpendicular lines passing through each of the apexes of the basic polygon. As with the Fischer projection, we do not show the carbon atoms forming a component of the basic chain, *i.e.* in this case the members of the ring. The Haworth projection is frequently used in the chemistry of carbohydrates in order to conveniently stress the cyclic character of the compounds. Therefore, we will give an example from this field, the formula of α-glucopyranose (*XXI*). As a rule, we usually also leave out the hydrogen atoms and show only the substituents different from hydrogen. The position of the oxygen atom, and thus also the position of the ring, are not prescribed, permitting the representation of the molecular model from the side best suited to the given purpose. The same method may also be used for other rings, with other atoms contained in the ring or with different sizes. In comparison with the Fischer projection, the distortion caused by the Haworth projection is less, even though the formula is still far from the actual shape of the molecule. A more exact picture of the molecule is given by conformational formulas only.

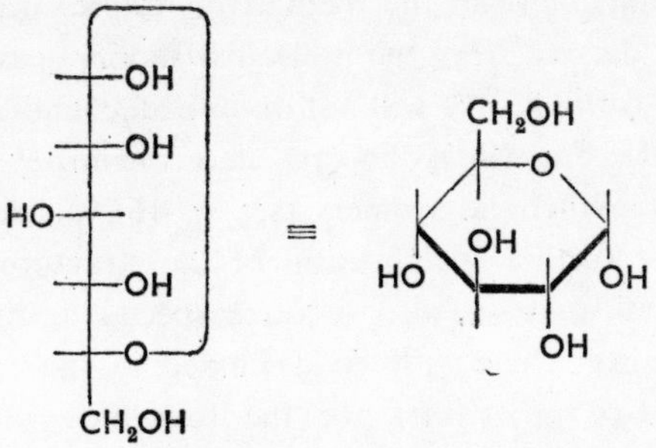

XXI

Although the Fischer and Haworth projections are generally applicable, they are encountered mainly in the chemistry of carbohydrates and amino acids. Similarly, the chemistry of steroids has a generally accepted special method of representing the molecule which is based on the almost planar character of the large steroid molecule. The polycyclic configuration of the steroid is usually projected into a plane parallel to the plane of the somewhat idealized steroid ring. The substituents attached to all of the skeletal atoms are divided into those located in front of the plane of the paper and those located behind it. The former are expressed by means of a heavy or solid line, the latter by a dashed connecting line. This method has spread from the chemistry of steroids into the closely related field of triterpenes and is now utilized in the chemistry of terpenes in general, as well as in the chemistry of alkaloids and other polycyclic natural compounds. In the case of relatively simple compounds, it often replaces the Fischer or Haworth conventions. However, even this method of representation, closely resembling the actual configuration, cannot replace conformational formulas if we attempt to reproduce the most probable appearance of molecule more truly.

Stereochemical Nomenclature

The naming of a compound is based on the structure of the molecule excluding the small number of trivial names. This generally accepted principle has to be maintained even in forming designations and symbols in the field of stereochemistry. It would not be expedient to attempt to derive a designation or configurational symbol from the origin of a compound or from the method of its preparation. The designation has to be derived from the molecular model or possibly from the correctly written formula. We will not discuss all chemical nomenclature, but only that part of it which concerns stereochemistry, *i.e.* the differentiation between geometrical isomers (see p. 16), diastereoisomers and enantiomers. The further specification of the structure, made possible by conformational analysis, also requires special designations for the individual positions. These will be discussed in the relevant chapter.

In principle, two procedures for the formation of stereochemical designations are possible. Either we designate the molecule as a whole by means of a symbol, or we provide each asymmetric atom (each element of chirality) in the molecule with a symbol and then designate the molecule with the help of a combination of the individual symbols. Both procedures are encountered in the literature.

The first stems from the fact that compounds cannot exist in more than two enantiomeric forms. Therefore, two symbols suffice for their designation: generally, one symbol for the configuration, which we may call positive, and one for the negative configuration. The procedure is in principle very simple but in the course of practical application we come up against two basic difficulties. The first appears with compounds having several asymmetric atoms, where several pairs of enantiomers appertain to one constitutional formula and thus also to one systematic designation. Therefore, it proved necessary to introduce a further designation besides the configurational symbol in order to distinguish the individual diastereoisomers. The second difficulty is in the ascribing of symbols to models of a certain type, *i.e.* in the general definition of the positive or negative configuration of the molecule.

The first problem, the differentiation of diastereoisomers, has been solved successfully. The basis of the designation is formed by the trivial names of aldoses with the respective number of asymmetric atoms, which, after the removal of the ending -se, provide the prefix designating the configuration. Thus the names of carbohydrates with two asymmetric atoms, erythrose and threose, provided the prefixes *erythro-* and *threo-*, designating the analogous configuration of compounds with two asymmetric atoms (formulas in the Fischer projection; see Diagram A 4). Thus, for example, we may designate mesotartaric acid systematically as *erythro*-2,3-dihydroxysuccinic acid and optically active or ra-

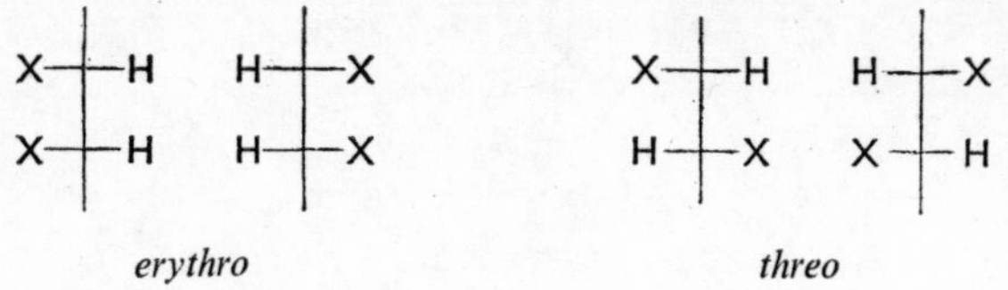

erythro *threo*

Diagram A 4

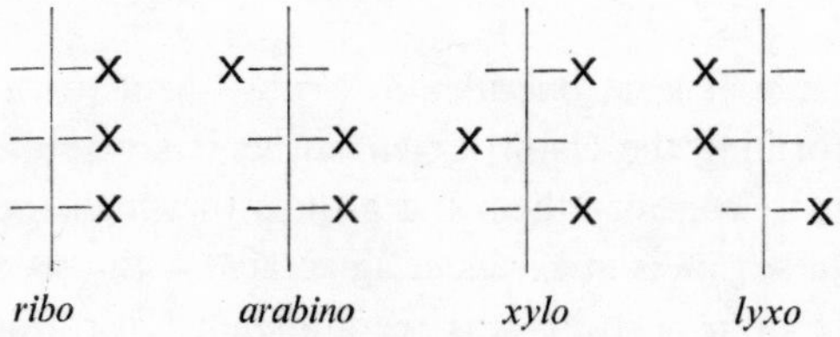

ribo *arabino* *xylo* *lyxo*

Diagram A 5

cemic tartaric acid as the respective *threo*-isomer. The prefixes for systems with three asymmetric atoms are derived from the names of the pentoses (Diagram A 5 shows the formulas of one of the enantiomers in each case).

The common position of the substituents of four asymmetric atoms is designated by means of analogous prefixes derived from the names of the hexoses (see Diagram A 6).

If the molecule incorporates a larger number of asymmetric atoms, the prefix for the four-atom system is combined with the prefix for the remaining number of asymmetric atoms. The prefixes may also be used for compounds in which the asymmetric atoms are not directly linked to one another. However, the prefixes are unequivocal only if all the asym-

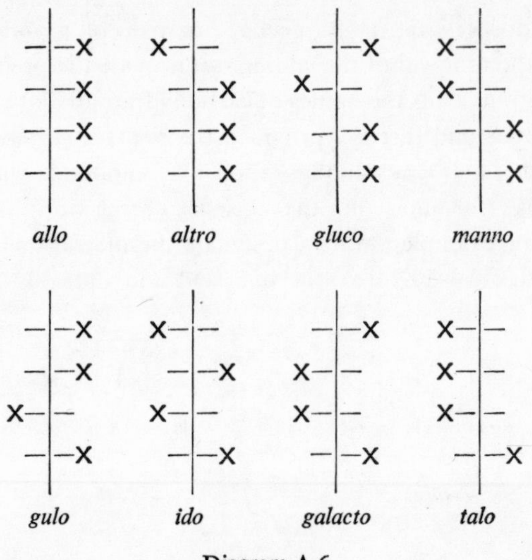

Diagram A 6

metric atoms have at least one identical substituent (with the exception of the atoms forming the chain) or substituents so similar in character as to enable us to consider them the groups to which the prefix refers. The structure of sugars is in excellent agreement with this condition, and therefore prefix nomenclature was recommended for this field in the proposals of the nomenclature commission of the International Union for Pure and Applied Chemistry (IUPAC). However, if two asymmetric atoms have absolutely different substituents or atoms, then this nomenclature cannot be used without additional explanation, as was the case with the *cis-trans* designation of ethylenes substituted in a different manner. Prefix nomenclature thus has inherently only limited validity. On the other hand, however, it has the great advantage of being applicable to either optically active compounds or to racemates and *meso*-forms, because it does not distinguish enantiomers but only diastereoisomers. A system designating each individual centre of chirality is rather cumbersome for optically inactive compounds.

The solution of the second difficulty with the nomenclature designating the molecule as a whole, *i.e.* the distinction of enantiomers, was less successful. The initial attempt to designate the configuration of enantiomers according to their sense of rotation was soon abandoned since it disagrees with the fundamental requirement of connecting the name and symbol with the structure (molecular model) and not with the substance itself. Moreover, the rotation of a compound depends on external influences, *i.e.* the solvent, wavelength, *etc.*, so that such a method would not be unequivocal.

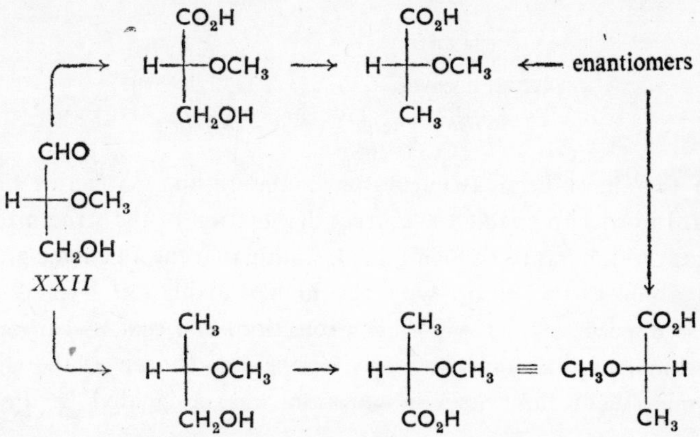

Diagram A 7

For a long time, then, symbols were ascribed in the so-called **natural series of substances.** A certain simple compound was conventionally designated by means of one symbol and its enantiomer by another. The series of compounds which were preparable from the chosen standard compound were then to bear the same designation. We may easily imagine that by means of simple operations we are sometimes able to prepare both enantiomers of another compound from the same standard, *e.g.* derivatives of both lactic acids from one glyceraldehyde derivative (*XXII*) (see Diagram A 7, formulas in the Fischer projection). Both products, as enantiomers, have to be designated by means of opposite symbols although, according to the "natural series", they should have the same symbol. The abundance of such examples prevented the formulation of a nomenclature system on the basis of chemical relationships. Names formally based on natural series are therefore applied in two specific regions of chemistry only, *i.e.* the chemistry of carbohydrates and the chemistry of amino acids. Carbohydrates are designated with the help of the symbol D or L according to whether their asymmetric atom with the highest ordinal number has, in the Fischer projection, a hydroxyl group on the same side as D- or L-glyceraldehyde, respectively.

CHO	CHO
H—+—OH	HO—+—H
CH_2OH	CH_2OH
D-glyceraldehyde	L-glyceraldehyde
XXIIIa	*XXIIIb*

These two have been assigned the configurations shown in Formula *XXIII*. In order to enable the correct designation of the structure, a further convention has to be maintained, stating that the aldehydic group or, in its absence, the group with the highest oxidation level is always written uppermost in the Fischer projection, and that its carbon atom has ordinal number one. Similarly, serine was chosen as the standard for amino acids and its configuration was designated by Formulas *XXIV*:

CO_2H	CO_2H
H—+—NH_2	H_2N—+—H
CH_2OH	CH_2OH
D-serine	L-serine
XXIVa	*XXIVb*

The other amino acids are designated by the symbols D or L according to whether the position of the amino group corresponds to that of D- or L-serine. The carboxyl group is located on top. The above designation of the configuration of carbohydrates and amino acids is internationally binding as it has been accepted by the IUPAC. The complete designation of a carbohydrate or amino acid is thus given by the configurational symbol (D or L), expressing its relationship with the D- or L-glyceraldehyde series, or to the D- or L-serine series; then by the configurational prefix chosen according to the number of asymmetric centres on the basis of the relative position of the functional groups of a compound in the Fischer projection; and finally by the systematic or trivial name (see for example Formulas *XXV–XXVII*).

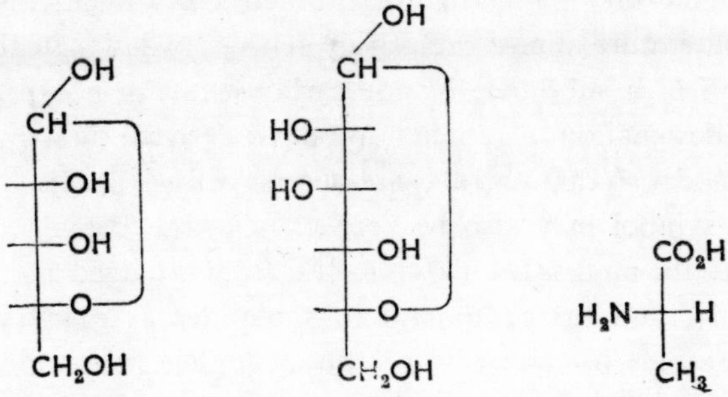

D-ribofuranose *XXV* D-mannopyranose *XXVI* L-alanine *XXVII*

In the case of amino acids with two asymmetric atoms we also encounter the prefix allo. This prefix designates an isomer prepared at a later date and thus has no direct relation to the configuration; it has the character of a trivial designation, not a systematic one. In the case of amino acids with several asymmetric atoms, if we intend to express the configuration in terms of this designation, we have to use the same nomenclature as with the carbohydrates. If a certain compound has simultaneously the character of an amino acid as well as a carbohydrate derivative, the configurational symbol (D, L) would have to be designated with the help of subscript $_g$ or $_s$, respectively, according to whether the configuration refers to glyceraldehyde or to serine (D_g, L_g or D_s, L_s).

Outside the regions of carbohydrate and amino acid chemistry this nomenclature cannot be applied consistently since a large number of not very lucid conventions would be necessary. First of all, the molecule would have to be orientated according to these conventions in order to permit the unequivocal choice of the Fischer projection; then a number of further conventions would have to determine which of the two substituents (right and left in the projection formula) is the one the position of which we are comparing with the hydroxyl group position in the glyceraldehyde or serine standards.

Consequently, several suggestions appeared, resolving the nomenclature and symbolism of chiral compounds generally by the second method, *i.e.* by designating the configuration of each centre of chirality independently. The simplest and most generally applicable of these suggestions is the Cahn-Ingold-Prelog one (6), the so-called ***R*/*S*-system**. It is now used in chemical literature almost exclusively in order to designate the absolute configuration of a substance of non-carbohydrate or non-peptide character. The designation of configuration is derived directly from the molecular model so that no rules are necessary for its projection. Nevertheless, the symbol may also be derived from the formula, if we know its relation to the model. Each asymmetric atom is judged and designated independently, and the compound thus requires as many symbols as asymmetric atoms (or elements of chirality). The procedure is similar for compounds where atropisomerism or other types of optical isomerism are encountered. Three conventions are necessary in order to designate the configuration of any centre of chirality whatsoever.

1. Determination of the order of substituents around the centre of chirality (by the term "substituent" we understand every atom or group of atoms directly attached to the centre, axis or plane of chirality).
2. The way of observing the model.
3. Determination of the symbol for the right and left sense of the sequence of substituents arranged in order.

For the determination of the seniority of substituents the authors chose the atomic number (not the weight) of the element directly attached to the asymmetric atom. If some of the atoms attached directly to the asymmetric atom are identical and the seniority of the substituents cannot be determined in this first order, the second or higher orders decide, *i.e.* the atomic number of atoms attached to the first-order atoms. The determination of the seniority of substituents for lactic acid (*XXVIII*) may serve as an example:

```
O     O—H
 \\  /
   C
   |
H—C—O—H
   |
   C
 / | \
H  H  H
```

XXVIII

In the first order, the atoms O, C, C and H are attached to the asymmetric carbon atom. The highest group therefore is the hydroxyl group; the hydrogen atom is last. The second and third place cannot be decided in the first order – both are occupied by carbon atoms. In one case (carboxyl group) oxygen atoms are attached to the carbon, in the other case (methyl group), hydrogen atoms are attached. Therefore, the carboxyl group is second and the methyl group third.

If the asymmetry of the molecule is caused by the different isotopes of one element only, then the isotope of larger atomic weight has priority over the isotope with the lesser one.

If two substituents attached to an asymmetric atom differ only as regards the configuration around a double bond, they cannot be distinguished on the basis of atomic numbers and we have to take into consideration a further rule according to which the *seqcis* isomer has priority over the *seqtrans* isomer. (A special designation of geometrical isomerism is used instead of the usual *cis-trans* one. The reference groups for this purpose are also chosen according to the rules governing the seniority of groups, and the designation need not always agree with the generally applied, but sometimes inconsequential, designation. In the instance of pseudoasymmetric atoms, which always have two substituents differing only as regards absolute configuration, priority must be given to the substituent with configuration *R* over the substituent with configuration *S* (the meaning of the symbols *R* and *S* will be explained later on). For optically active compounds having no asymmetric atom (allenes, atropisomers, *etc.*), one rule has priority over all rules concerning seniority – the rule stating that the nearer substituent enjoys priority over the more distant one. In such cases, namely, pairs of more distant substituents may be identical (see p. 33) and, according to the previous priority rules, no decision between them could be made.

Because of the considerable importance of this convention, we will summarize all the necessary rules once more:

(1) The nearer end of the axis and or the nearer side of the plane has priority over the more distant one.

(2) The higher atomic number of an atom has priority over the lower one.

(3) The higher atomic weight enjoys priority over the lower.
(4) *Seqcis* enjoys priority over *seqtrans*.
(5) The following configurations have priority:
R, *R* or *S*, *S* over *R*, *S* or *S*, *R*; *M*, *M*, or *P*, *P* over *M*, *P* or *P*, *M*; *R*, *M* or *S*, *P* over *R*, *P* or *S*, *M*; *M*, *R* or *P*, *S* over *M*, *S* or *P*, *R*; *r* over *s*.
(6) *R* has priority over *S* and *M* over *P*.
(The meaning of the symbols *R*, *S*, *M*, and *P* is explained below.)

The rules have to be employed in the order in which they are listed and each rule may be applied only if we are unable to arrive at a decision on the basis of the preceding rule.

When the order has been determined, we orientate the molecule in such a manner as to be able to ascribe the correct symbol. According to the author's suggestion, the molecule ought to be viewed from the side most distant from the lowest substituent, in the direction of its bond. The order of the first three substituents then has the clockwise or anti-clockwise sense. The clockwise sense is designated by the letter *R*, the opposite sense by the letter *S*; analogously *r* or *s* are used for a pseudo-asymmetric atom. It is easier to understand if one imagines that the bond connected to the lowest substituent forms the shaft of a steering wheel, on the periphery of which the three substituents of higher order are symmetrically distributed (Fig. 4). If their sequence indicates the sense of rotation of the steering wheel to the right, then the asymmetric atom is identified by the symbol *R*, whereas *S* indicates rotation to the left (from the latin words *rectus* – right, *sinister* – left). The authors deliberately chose signs different from the usual D and L in order to emphasize the difference in the meaning of these symbols written with the ordinal number of the asymmetric atom to which they refer, in front of the whole designation of the compound.

Sometimes it is better, or even necessary, to consider chirality as a helix; in this case the right-handed helix is designated by the symbol *P* (plus), the left-handed one by the symbol *M* (minus) (see p. 31). Symbols other than *R* and *S* were chosen in order to indicate directly the method of observation of the model, which may sometimes be described by the configurational (*R*, *S*) symbols or by the helicity symbols (*M*, *P*).

Considering the growing importance of this method of designating configuration, it is advisable to study its principles in greater detail in the original literature (6) or in a summarizing report (7).

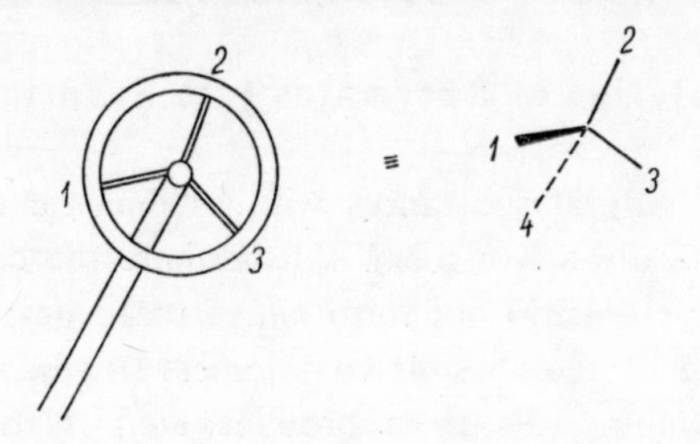

Fig. 4. Determination of the order of substituents according to Cahn, Ingold and Prelog. Whilst D and L designate a configuration identical with the configuration of D- or L-glyceraldehyde or serine, respectively, the signs *R* and *S* designate the absolute configuration around one asymmetric atom of an arbitrary molecule.

Finally, we have to consider the stereochemical meaning of the often-used symbols α and β. They appear with two meanings. In the chemistry of carbohydrates they indicate the position of the substituent other than hydrogen attached to the first (aldehydic) carbon atom. The definition has changed somewhat from time to time but today the generally accepted α-configuration is the position identical with the configuration around the carbon atom determining the pertinence of the molecule to the D- or L-series (*i.e.* the configuration around the asymmetric atom with the highest ordinal number), whereas β applies to the opposite configuration. In the chemistry of steroids, α designates the position of the substituents under the plane of the paper, *i.e.* behind the plane of the steroid skeleton, whereas β designates the position above the plane of the paper. Since in the normal steroid nucleus the methyl group attached to $C_{(10)}$ is situated above the plane of the paper (β), all β-substituents have a *cis*-relationship to this methyl group. In the usual steroid projection, α-substituents are thus designated by means of a dashed line and β-substituents are attached by means of bonds drawn as full lines. The α- and β-designation with the same meaning is today becoming routine for a number of other natural polycyclic compounds, expecially where an analogous method of projection for the writing of formulas had penetrated.

Resolution of Racemates into Enantiomers

The majority of natural substances with asymmetric centres exist in nature in an optically active form. Furthermore the compounds with more asymmetric atoms do not form native racemates; usually we encounter only one of the possible enantiomers (D-glucose, D-galactose, L-arabinose, L-amino acids from proteins, *etc.*). Natural substances are formed by highly stereospecific biological processes. As a rule, synthesis in the laboratory yields the asymmetric compounds in the form of racemates. An optically active compound is obtained directly only if the parent substance is optically active itself. Isolated cases of synthesis of an optically active compound from a symmetrical or racemic compound without the assistance of enzymes are, as yet, preparatively of little value (see chapter on asymmetric synthesis).

Thus, we often have to perform the task of resolving a racemic compound into its optically active components. Physical, chemical and biological methods have been worked out for the resolution of racemates. As we have already stated (p. 21), racemates are sometimes physical mixtures of the same quantities of enantiomers but are more often a molecular compound. We are therefore only very seldom able to separate a racemate into its components by physical means, *e.g.* by the mechanical sorting of crystals. Although the very first artificial resolution of a racemate was carried out in this manner, this method of separation has no practical importance.

The great majority of preparative resolutions of racemates have until now been effected by chemical means. The principle of the method is the introduction of a new centre of asymmetry into the molecule by means of synthesis with an auxiliary, optically active substance. Thus the racemate yields two diastereoisomeric compounds which may be separated by the usual separation methods, usually by crystallization. After the separation of the diastereoisomers, the auxiliary compound is removed and the initial substance liberated in the optically active form. Schematically, we may illustrate the procedure as follows:

$$(\pm)A + (+)B \longrightarrow \begin{array}{ccc} (+A\text{—}+B) & + & (-A\text{—}+B) \\ \downarrow & & \downarrow \\ (+)A + (+)B & & (-)A + (+)B \end{array}$$

$$(\pm)A + (-)B \longrightarrow \begin{array}{ccc} (-A\text{—}-B) & + & (+A\text{—}-B) \\ \downarrow & & \downarrow \\ (-)A + (-)B & & (+)A + (-)B \end{array}$$

where $(\pm)A$ is the racemate to be resolved and $(+)B$ and $(-)B$ are the optically active auxiliary compounds. It goes without saying that we have to have some compound in an optically active form, usually a natural substance. Our task lies in finding such a compound B which would make the diastereoisomers of compound A—B markedly different as regards their physical properties, *e.g.* their solubility. The ideal case is if one diastereoisomer is insoluble and the other one well soluble in the chosen solvent so that after the combination of components A and B one diastereoisomeric product is precipitated and the other remains in solution. Usually, however, the diastereoisomers are only slightly different and repeated recrystallization is necessary to obtain one of them in pure condition. The more soluble diastereoisomer (and thus also the second enantiomer of substance A) can normally be purified only with extreme difficulty. The second enantiomer is recovered more readily by using another auxiliary substance, preferably the enantiomer of the substance originally applied. For example, if compound $(+A\text{—}+B)$ of the two possible diastereoisomers crystallizes better than compound $(-)A + (+)B$, then from the opposite pair of isomers compound $(-A\text{—}-B)$ must crystallize better than compound $(+A\text{—}-B)$.

The most frequent reaction of this type is the formation of a salt from a racemic base with an optically active acid, or from a racemic acid with an optically active organic base. Racemic acids are mostly resolved with the help of alkaloids, *e.g.* cinchonine, quinine, quinidine, cinchonidine, brucine, strychnine, morphine, ephedrine, but α-phenylethylamine, 1-phenyl-2-amino-1,3-propanediol and a number of others have also been used. On the other hand, organic bases are often resolved by means of tartaric acid and its derivatives, *e.g.* dibenzoyltartaric acid, ditoluene-*p*-sulphonyltartaric acid *etc.*, as well as malic acid, mandelic acid, sulphonic acids derived from camphor, dinitrodiphenic acid, *etc.* An amino acid may also be resolved as an acid if its amino group is protected by suitable acylation. Phthaloyl derivatives prove especially successful. On the other hand, esters of some amino acids have been successfully resolved as bases with the help of optically active acids. Alcohols are converted, by esterification with phthalic anhydride, into acid phthalates and in this form resolved as acids by means of optically active bases. Similarily, ketones can be converted by means of a reaction with semicarbazideacetic acid into derivatives incorporating a carboxyl group capable of forming salts.

Acids and alcohols are seldom resolved in the form of esters with an optically active component. The esters of carboxylic acids crystallize in a poor manner, and attempts to separate the diastereoisomeric esters by distillation have seldom been successful. The formation of amides and similar derivatives for the chemical resolution of racemates also remains practically unused, mainly because of the difficulties connected with the regeneration of optically active resolution products from the intermediate product.

A more difficult situation arises with compounds which have no functional groups. In the case of aromatic hydrocarbons, use can be made of their ability to form complexes with polynitro compounds. By the introduction of a carboxylic group into a polynitro derivative, we may form a stable complex allowing further separation. Even an olefin of the alicyclic series has recently been resolved into its enantiomers on the basis of the formation of a complex compound with a palladium derivative incorporating an asymmetric organic base. Saturated hydrocarbons have as yet not been chemically resolved with success.

However, as a reaction with an optically active substance leads to an optically active compound – if no side reactions or elimination of molecular asymmetry occur – in the same way as a racemate leads to a racemate, the difficulties of resolving racemates without functional groups are circumvented in such a manner that the resolution is carried out at some other stage of the synthesis and the latter is then pursued further, already with optically active material.

The stage at which we wish to effect the resolution may be chosen within broad limits and the best method always has to be determined experimentally.

Although chemical resolution is applied most frequently, it is far from perfect, as it usually necessitates a very painstaking und uneconomic separation of rather similar diastereoisomers. There is no generally applicable recipe for the execution of the operation and the yields are seldom satisfactory. It is therefore very advisable to seek further racemate resolution procedures. A number of them have already been described in the literature; as yet, however, none of the methods is generally applicable. The methods which could, in due course, gain practical importance will be referred to.

Special attention has recently been given to chromatography because of its high separating efficiency. In principle, we are able to follow two paths when applying chromatography to the resolution of racemates. The first one is a modification of the resolution method such that the diastereoisomers are not separated by crystallization but by means of chromatography, especially preparative vapour phase chromatography. As the intermediate diastereoisomers are usually salts, this application of chromatography is very limited. The chromatography of diastereoisomeric esters or amides seems to be promising.

The second method, *i.e.* the resolution of racemates by means of chromatography on an optically active carrier, has been successful in individual cases. In this procedure it is not necessary to prepare the two diastereoisomers from the racemate and from the auxiliary optically active compound, because the two enantiomers of the racemate may be attached to a carrier of chiral structure with different strength, and thus separated. Favourable results have been obtained with lactose and starch as carriers. Also, paper may be considered an asymmetrical carrier and actually a number of cases have been described in which partial resolution of a racemate occurred during paper chromatography. It is also possible to adsorb the racemic compound on to an achiral carrier, *e.g.* silica gel, and then to carry out elution chromatography of enantiomers with a chiral eluent. However, much experimental work has yet to be done in this field. Best-known are the conditions of resolution on starch, for which it is even possible to formulate several principles of a more general nature. First of all, the racemate has to be soluble in water or at least in aqueous ethanol or aqueous acetone. The resolved molecule has to have more than one polar group to enable it to be attached to the carrier at more than one point. Only thus can the asymmetric structure of starch assert itself. It is advantageous if one of the polar groups is attached directly to the asymmetric atom. Amino acids are better resolved in the form of benzoyl derivatives.

Still less is known of the resolution of racemates with the use of ion exchangers. This procedure is just beginning but its application will probably spread. Until now, the resolution method was such that an optically active compound (for instance tyrosine) was adsorbed on the exchanger and the racemate was chromatographically processed on the treated column. A more effective method is the direct chemical modification of the ion exchanger, for instance by the reaction with thionyl chloride and esterification of the intermediary with quinine. Amberlite treated in this manner was succesfully applied to the resolution of DL-mandelic acid.

Chromatography on an optically active carrier is similar to crystallization from an optically active solvent which also does not require preliminary treatment of the racemate. This method was successful in partially resolving a racemic compound, as one of the enantiomers was more soluble in the chiral solvent than the other. Of more practical interest is the resolution of racemates by spontaneous crystallization or, in other words, crystallization from supersaturated solutions. The process is based on the fact that after the addition of one enantiomer to a racemic solution of the same compound, a larger quantity of active material crystallizes out of the solution than has been added. Although the method is applicable for some compounds only it has great technical advantages. By repeating the process it is possible to obtain both pure enantiomers simply by crystallization. The method is technically utilized for the preparation of amino acids.

Spontaneous resolution into enantiomers was also encountered in the formation of addition compounds with urea. Urea clathrates appear in two antisymmetric forms. In especially favourable cases one of them may crystallize out first. The preparation of the optically active substance is easier if a pure enantiomer is available for seeding the solution. This method is, however, also not generally applicable; according to literature data it was used to resolve the asymmetric hydrocarbon 3-methyloctane, which is unresolvable by other means.

Very important, and preparatively as well as technically significant, is the old method of resolution based on enzymatic reactions. The first procedure, Pasteur's, is based on the finding that some micro-organisms utilize for their nutrition only one enantiomer of a certain substance. For example, if a culture of a certain mould is grown on the racemate solution and if its growth is interrupted after a suitable period, the remaining solution contains a more or less pure, optically active component of the parent substance, namely the component not utilized by the culture for its nutrition. The procedure is rather exacting experimentally and it is always necessary to determine the suitable cultivation time because the mould, after consuming the favoured enantiomer, usually starts to consume the other one and all the material would thus disappear from the solution after a longer period of time. The second procedure utilizes the specific enzymatic reactions directly. As a rule, enzymes catalyse reactions of a certain type with substances of one configuration only. Thus, if a racemate is treated with an enzyme, only one component will react and the second one will remain unchanged so that in a favourable case it will be possible to isolate both (one changed, the second one unchanged). For example, if we treat a racemic α-amino acid amide in solution with the enzyme L-leucineamidopeptidase, only the L-amino-acid amide will be hydrolysed and free L-amino acid will appear in the solution together with the unhydrolysed D-amino acid amide. The reaction products can then be relatively easily separated, for instance on the basis of salt formation. Contrary to all the other methods of racemate resolution, this method will directly determine the configuration of the isolated optically active compound. In working with an isolated enzyme, there is moreover no danger of overdoing the reaction, which can happen with live cultures. Cases have been described, when the method succeeded in producing more of one enantiomer than is "theoretically" possible, *i.e.* more than 50 % of the racemate.

If, in the stated example of amino acid amide decomposition, the parent amide were an optically unstable substance, rapidly racemizing in the reaction medium, whereas the liberated acid were optically stable, then as the enzyme decomposes one enantiomer (L) the second enantiomer, remaining in the mixture, racemizes, *i.e.* is converted into the L-configuration until equilibrium is attained, and is transformed into the stable L-acid and so on until practically all the amide is converted into one configuration of the free amino acid. Although such a case of optical transformation is also possible with chemical resolution, *i.e.* during the resolution of relatively unstable, asymmetric β-aminoketones, it is more usual in the field of enzymatic reactions.

The disadvantage of enzymatic procedures is the considerable lack of experience with work of this type on the part of most organic chemists. Nevertheless, the enzymatic resolution of racemates will without doubt soon occupy the first place in this field of chemistry as it is simple in comparison with all the other methods and its yields may be practically quantitative. The reaction is limited in that it can be carried out only with compounds very analogous to natural substances, as the enzymes required for substances not appearing in nature are not available. In the meantime we are also confined by the insufficient availability of an adequate range of pure cultures or even pure enzymes.

NOTE. A novel system for designation of configuration of geometrical isomers (8) is now used by Chemical Abstracts. A decision is made as to which of the two groups attached to each of the doubly bound atoms has the higher priority using the sequence rules (see p. 55) and then the configuration in which the two groups of higher priority are on the same side of the "double bond plane" is denoted by the prefix Z (zusammen) and that in which groups are on opposite sides is given the prefix E (entgegen).

Literature

1. Daudel R.: *Electronic Structure of Molecules*. Pergamon Press, Oxford 1966.
2. Lide D. R.: Tetrahedron *17*, 125 (1962).
3. Svoboda M., Sicher J.: Chem. and Ind. (London) *1959*, 290.
4. Cope A. C., Howell C. F., Knowles A.: J. Am. Chem. Soc. *84*, 3190 (1962).
5. Briegleb G. in: *Houben-Weyl, Methoden der organischen. chemie* Band 3/1, *Physikalische Forschungsmethoden* (E. Müller, Ed.), p. 545. Thieme Verlag, Stuttgart 1955.
6. Cahn R. S., Ingold C. K., Prelog V.: Experientia *12*, 81 (1956); Angew. Chem. *78*, 413 (1966).
7. Cahn R. S.: J. Chem. Educ. *41*, 116 (1964).
8. Blackwood J. E., *et al.*: J. Am. Chem. Soc. *90*, 509 (1968).

B. DETERMINATION OF RELATIVE AND ABSOLUTE CONFIGURATION

Chemical Procedures for Determination of Absolute Configuration

A number of chemical reactions are often necessary to enable us to ascribe unequivocally a constitutional formula to a certain compound. Moreover, the correct spatial formulas have to be assigned to diastereoisomers and enantiomers. The determination of the relationship between the substance – the chemical individual – and the spatial formula or molecular model, respectively, is usually designated as the determination of the relative or absolute configuration.

Correlation

Until approximately the second half of this century it was possible to determine the configuration of a compound only in relation to a similar compound which had been chosen as a standard. Hence the term "correlation", *i.e.* assignment, connection. The fundamental, standard compounds were assigned a configuration purely arbitrarily, on the basis of agreement, and the configuration of the other compounds was correct only in relation to the chosen standard. For various types of compounds, namely for various types of asymmetric carbon atoms, it proved necessary to choose various standard compounds. Thus, several correlation series originated, and, for a long period, there was no connection between them. The most important standard compound is glyceraldehyde, the (+)-enantiomer of which was assigned the configuration as shown in Formula *XXIIIa*. Glyceraldehyde forms the basis of correlation for all carbohydrate compounds first of all and for a series of further compounds with an asymmetric atom of the secondary carbinol type. Similarly, the standard for natural amino acids is the (+)-enantiomer of serine, which was assigned the analogous formula *XXIVa.* Standard compounds with an arbitrarily determined configuration were selected in the same manner for some other types of optically active compounds. The configuration, determined by forming a chemical relationship (correlation) of the compound to the standard, was correct only if the configuration of the standard was also correct. It was quite as probable that all spatial formulas represented the enantiomers of the compounds to which they were assigned. Fortunately, it proved (1) that the chosen configurations of glyceraldehyde, and consequently also the configurations of serine, represent the true spatial arrangement, so that all the formulas defined by their relationship to these basic compounds express the absolute configuration of the compound directly.

The basic requirement for correlation by means of chemical reactions is that no reaction be permitted to attack the asymmetric atom directly or indirectly. A determination of the configuration complying with this basic requirement is generally accepted as fully valid. In order to execute chemical correlation we may thus use all the reactions performable without damage to the structure attached to the asymmetric atom. The greater the stability of the asymmetric atom, the larger the number of changes we may carry out in another part of the molecule. Series of compounds with asymmetric atoms of the same type are formed by means of chemical correlation defined in the above manner. Several simple examples of effected correlations will now be discussed.

CHO / H—|—OH / CH_2OH (+) glyceraldehyde $\xrightarrow{HgO}$ CO_2H / H—|—OH / CH_2OH (−) glyceric acid $\xleftarrow{HNO_2}$

$\xleftarrow{HNO_2}$ CO_2H / H—|—OH / CH_2NH_2 (+) isoserine $\xrightarrow[\text{2. NaHg}_x]{\text{1. NOBr}}$ CO_2H / H—|—OH / CH_3 (−) lactic acid

Diagram B 1

For example, the configuration of glyceric acid formed by oxidation may be derived directly from the standard glyceraldehyde. The acid may then be related in two steps with lactic acid, thus also determining the configuration of the latter (Diagram B 1). The configuration of racemic tartaric acid is determined for instance by the cyanohydrin synthesis of the left-handed enantiomer from D-glyceraldehyde. β-Hydroxybutyric acid may be prepared from lactic acid, the configuration of which is already known to us. The former may be reduced, in four steps, to optically active 2-butanol, the most simple asymmetric alcohol (see Diagram B 2).

COOH / H—C—OH / CH_3 (—) $\xrightarrow{Na}$ CH_2OH / H—C—OH / CH_3 (—) $\xrightarrow{HBr}$ CH_2Br / H—C—OH / CH_3 $\xrightarrow{KCN}$

$\xrightarrow{KCN}$ CH_2CN / H—C—OH / CH_3 (—) $\longrightarrow$ CH_2CO_2H / H—C—OH / CH_3 $\longrightarrow$

$\longrightarrow$ CH_2CH_2OH / H—C—OH / CH_3 (—) $\longrightarrow$ CH_2CH_2I / И—C—OH / CH_3 (—) $\xrightarrow{H_2/Pd}$ CH_2CH_3 / H—C—OH / CH_3 (—)

Diagram B 2

The indicated method was used for the determination of the configuration of the end-products as well as that of all the isolated and characterized intermediates. By means of analogous procedures we may correlate almost all the compounds with a hydroxyl group attached to the asymmetric carbon atom, namely carbohydrate compounds.

The direction of the synthesis is irrelevant for the correlation. We may start from a compound of known configuration and synthesize a compound whose configuration we are determining as well as the other way round. We even often proceed in both directions simultaneously by preparing a third compound from the known as well as from the determined compound. The configuration may be determined with the same reliability even if the two-way synthesis does not produce identical compounds but enantiomers. Correlation in the field of organic bases and amino acids may serve as examples (see Diagram B 3).

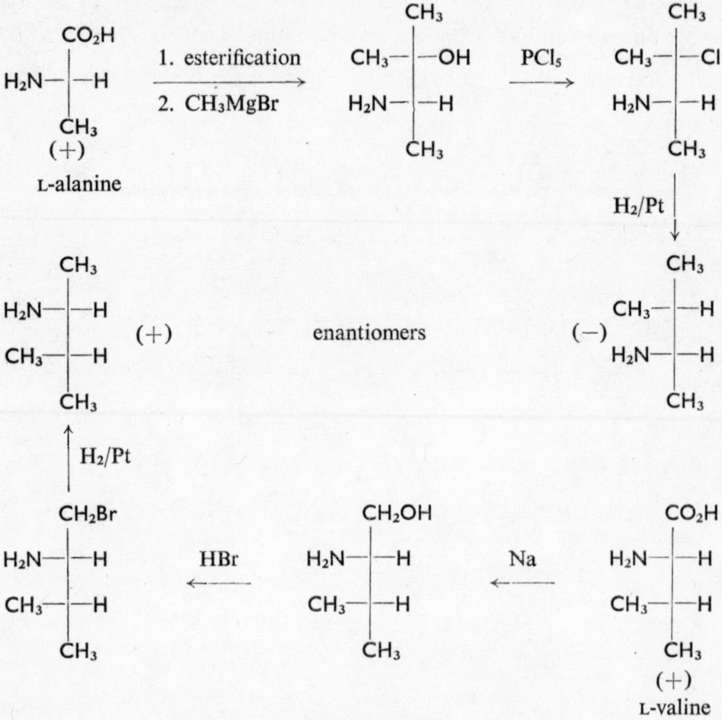

CO_2H
H_2N—|—H
CH_3
(+)
L-alanine

1. esterification
2. CH_3MgBr
→

CH_3
CH_3—|—OH
H_2N—|—H
CH_3

PCl_5
→

CH_3
CH_3—|—Cl
H_2N—|—H
CH_3

H_2/Pt
↓

CH_3
H_2N—|—H
CH_3—|—H
CH_3
(+)

enantiomers

(−)
CH_3
CH_3—|—H
H_2N—|—H
CH_3

↑
H_2/Pt

CH_2Br
H_2N—|—H
CH_3—|—H
CH_3

HBr
←

CH_2OH
H_2N—|—H
CH_3—|—H
CH_3

Na
←

CO_2H
H_2N—|—H
CH_3—|—H
CH_3
(+)
L-valine

Diagram B 3

An interesting example was the correlation between aliphatic and aromatic amino compounds which served to determine the configuration of phenylethylamine and related compounds (see Diagram B 4). The example shows that L-alanine and D-phenylglycine may easily be directly related. Correlation always requires exact knowledge of the reactions used for the correlation; configurations may not be assigned on the basis of nomenclatural symbols only. This is one of the many testimonies

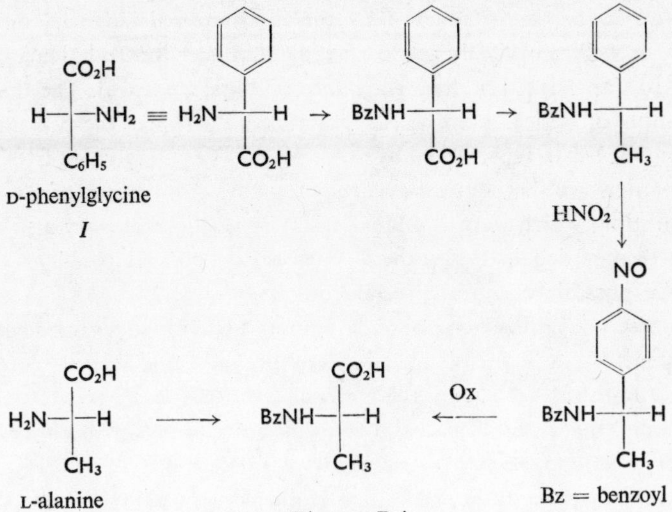

CO_2H / H—C—NH_2 / C_6H_5 ≡ H_2N—C—H (C_6H_5 above, CO_2H below) → BzNH—C—H (C_6H_5 above, CO_2H below) → BzNH—C—H (C_6H_5 above, CH_3 below)

D-phenylglycine

I

HNO_2 ↓

BzNH—C—H (p-nitrosophenyl, NO, above; CH_3 below)

CO_2H / H_2N—C—H / CH_3 ⟶ CO_2H / BzNH—C—H / CH_3 ⟵(Ox) BzNH—C—H (p-nitrosophenyl above; CH_3 below)

L-alanine

Bz = benzoyl

Diagram B 4

to the fact that nomenclatural symbols cannot be related to the origin of a substance and have to be connected therefore only with its structure.

The chemistry of terpenes and steroids presents another type of asymmetric atom, namely the tertiary carbon atom (branched carbon chain). As the standard in this series we may choose the simplest representative, fermented optically active amyl alcohol, which has been correlated, by means of relatively intricate procedures involving methyladipic acid, with many terpenes. Almost all the more usual terpenoid compounds are today correlated between themselves.

Another series has been formed by the reactions of optically active biphenyl derivatives, *e.g.* by the reduction of the nitro groups in dinitrodiphenic acid to amino groups, by the reactions of carboxyl groups, *etc.*

The basic principle of chemical correlation, prohibiting the execution of reactions directly at the site of asymmetry, however, prevents the analogous formation of relationships between the individual series just described. On the basis of chemical correlation we are therefore *a priori* forced to define a standard compound for each type of asymmetric compound. We have just seen examples of four standards, *i.e.* glyceraldehyde and serine, accepted even by the international chemical organisations, as well as optically active amyl alcohol and dinitrodiphenic acid, which so far have not had such international approval. The mutual correlation of series, namely the correlation between these standard compounds, has to be achieved by other means. Very important from this point of view are the physical and enzymatic determinations of absolute configuration, which we will discuss later; synthetic chemistry also need not be disregarded in this respect. With regard to this problem, there are three possible chemical procedures.

The first one utilizes compounds containing both investigated types of asymmetry (we will call them key compounds); the relative configuration of both substituents in the molecule *(erythro-threo)* is determined, and finally the usual classical chemical correlation of both centres with the individual respective standards is carried out.

The second procedure utilizes contemporary knowledge of the steric course of some substitution reactions and on this basis determines the configuration of the product formed by the direct substitution of one functional group for another one. Correlation between the glyceraldehyde and serine series is greatly facilitated by knowledge of the stereochemical course of the substitution of the toluene-*p*-sulphonyloxy group by the azide group:

$$R{-}OTs + N_3^- \longrightarrow R{-}N_3 + OTs^-$$

The toluene-*p*-sulphonyl ester is obtained from the alcohol and the azide may be converted to an amine by reduction, also without change of the configuration. Reactions of this type will be discussed in greater detail later (p. 142).

The third procedure is based on the utilization of the known course of an asymmetric synthesis; for example the absolute configuration of the series of biphenyl atropisomers was determined according to this procedure, thus forming a correlation between biphenyl derivatives and the glyceraldehyde series. This procedure will also be discussed in the chapter on asymmetric synthesis (p. 225). Therefore, let us return now to the first procedure, which is nearest to classical chemical correlation.

The relation between glyceraldehyde and serine was determined with the help of the above procedure by means of a derivative of D-glucosamine as the key compound. The conversion of D-glucosamine into D-glucose or D-mannose with nitrous acid established the configuration of the hydroxyl groups with regard to glyceraldehyde. The relationship between the asymmetric carbon attached to the amino group and the serine was determined by means of the oxidation of methyl 2-acetamido-2-deoxy-α-D-glucopyranoside by periodic acid and by further oxidation of the aldehydic groups to carboxyl groups and by the reduction of the original aldehydic function after the cleavage of the acetal group (see Diagram B 5). The relationship thus established confirmed once again the configuration previously arbitrarily assigned to serine in relation to the configuration of glyceraldehyde, also assigned arbitrarily.

Determination of Relative Configuration

Nowadays, if we know the absolute spatial structure of standard compounds, it is, as follows from the example just introduced, possible in principle to determine the configuration of substances with

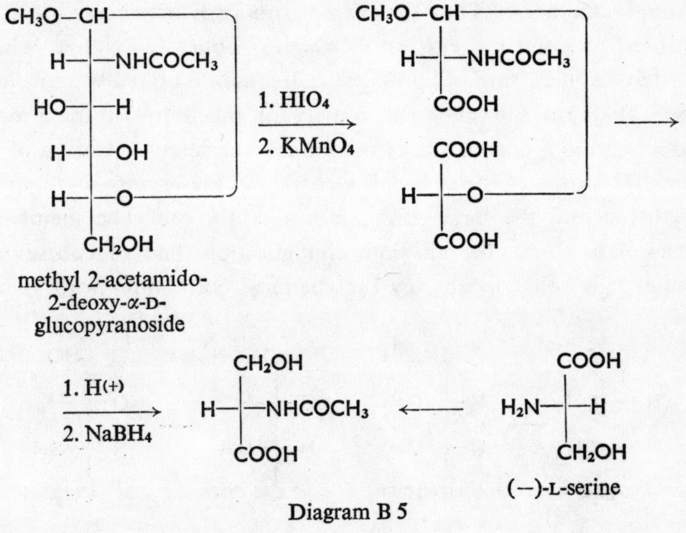

CH_3O-CH
H—|—$NHCOCH_3$
HO—|—H
H—|—OH
H—|—O
CH_2OH
methyl 2-acetamido-2-deoxy-α-D-glucopyranoside

1. HIO_4 → 2. $KMnO_4$

CH_3O-CH
H—|—$NHCOCH_3$
COOH
COOH
H—|—O
COOH

→

1. H(+) → 2. $NaBH_4$

CH_2OH
H—|—$NHCOCH_3$
COOH

←

COOH
H_2N—|—H
CH_2OH
(—)-L-serine

Diagram B 5

several centres of chirality by means of chemical correlation. Thus also we may determine the mutual relationship of these centres in the individual diastereoisomers, the so-called **relative configuration** (for example the *erythro*- and *threo*- configurations, respectively, of compounds with two asymmetric atoms). Usually, however, simpler procedures are used for the determination of the mutual relationship of several centres in the same molecule. Such methods determine the distribution of substituents in the molecule with relative reliability, and thus also allow the formulation of relationships between the individual configurational series.

With certainty we may determine the relative configuration of compounds with two identical asymmetric atoms. We already know, from the analysis of the conditions of optical activity (p. 49), that the inactive *meso*-form of a compound with two asymmetric atoms is bound to have the *erythro*-configuration, and the second diastereoisomer therefore has to be the *threo*-form. If we succeed in resolving one of the two synthetic diastereoisomers of this type into optically active compounds, this resolvable isomer has to have the *threo*-configuration, and the second one the *erythro*-configuration. The same principle may be applied to compounds into which a plane of symmetry may be introduced by means of a simple operation. Thus for example optically active erythrose may be reduced to a tetritol (erythritol) which is optically inactive, whereas its diastereoisomer, threose, provides on reduction optically active threitol (see Diagram B 6). Optical activity or inactivity of the products of reduction quite unequivocally determines the relative relation of both asymmetric atoms of the original tetroses, *i.e.* the *erythro*-configuration of erythrose and the *threo*-configuration of threose. The membership to series D or L, *i.e.* the absolute configuration, had of course to be determined by other means, by the chemical correlation with glyceral-

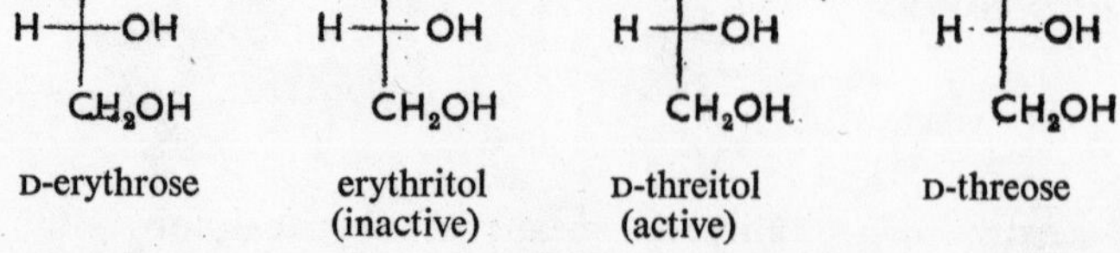

CHO
H—OH
H—OH
CH_2OH

→

CH_2OH
H—OH
H—OH
CH_2OH

CH_2OH
HO—H
H—OH
CH_2OH

←

CHO
HO—H
H—OH
CH_2OH

D-erythrose — erythritol (inactive) — D-threitol (active) — D-threose

Diagram B 6

dehyde or by physical means. Compounds with identical asymmetric atoms naturally cannot be key compounds and therefore this simple procedure may not be used to determine the relationship between correlation series; other methods are required.

The chemical methods of determining relative configuration are based on establishing the difference in the course of reactions according to a known mechanism for both diastereoisomers. From the aspect of evaluation the simplest are those which bring about the formation of a ring containing both asymmetric atoms (2). The configuration is assessed from stability, ease of formation and, in the most favourable case, from the possibility or impossibility of formation when comparing both diastereoisomers. The rings considered can be either reaction products or even intermediates assumed with sufficient reliability. Among the stable rings are aminoalcoholic complexes with cuprous, cobaltous or borate ions, derivatives with aldehydes or ketones (*e.g.* cyclic oxazolidines or tetrahydro-oxazines); transient ring formation is assumed for the migration of acyl groups, oxidation by periodic acid *etc.* A hydrogen bond also closes a ring between two functional groups attached to asymmetric atoms; its stability, which may be measured for example by means of IR spectra (3) or with the help of dissociation constants, may determine the configuration of the substance (4).

The heterocyclization reactions of aminoalcohols with aldehydes or ketones are valuable methods because the rings formed are stable and the cyclization products may thus be isolated. In some cases, however,

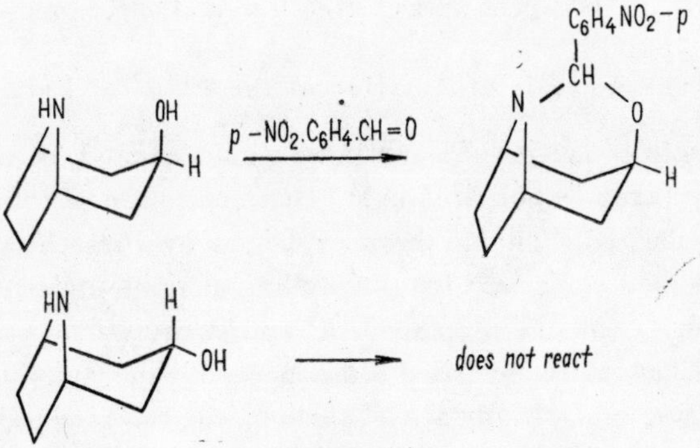

Diagram B 7

one diastereoisomer forms a cyclic derivative with the aldehyde, whereas the second one does not. This happens for example with the reaction of *p*-nitrobenzaldehyde with α- and β-nortropine (see Diagram B 7). As a rule, both diasteroisomers react, but the stability of the reaction products as regards ring re-opening differs. On the basis of conformational analysis it is then possible to determine the relative configuration of amino-alcohols and dihydroxy compounds using similar reactions (see p. 126).

Methods Based on Optical Rotation

The rotation of the plane of polarized light, the most characteristic property of optically active substances, forms the basis of a number of methods of determining absolute configuration (5,6). The simple rules based on the comparison of the rotation of compounds which have similar structures and are correlated by chemical means may of course be applied only to a limited extent and with caution. Such procedures usually are sufficiently reliable only within the scope of the homologous series of compounds with a single asymmetric atom. Their extension to more complex compounds, especially to compounds with a larger number of elements of chirality, based on the validity of the so-called **principle of superposition**, *i.e.* on the assumption that the rotation increments of the individual centres and groups are mutually independent and additive, proved to be unjustified. Results of a more generally applicable nature were provided only by the recognition of the relationship between optical rotation and electronic absorption.

Physical Fundamentals of Rotation of the Plane of Polarized Light

By the optical activity of a substance we understand the capability of rotating the plane of polarized light. The mechanism of rotation of the plane of polarized light has been explained by Fresnel.

A ray of plane polarized light, travelling at right angles to the plane of the paper, oscillates in plane *AB*, and is equivalent to the sum of right and left circularly polarized components of equal amplitude (Fig. 5). When entering an optically active medium, the two circularly polarized rays travel on at different velocities. At the exit from the optically active medium, the rays occupy two positions, A_R and A_L. After the two rays rejoin, the light now oscillates in plane $A'B'$, deflected from the original plane AB by the angle α. A plane polarized ray may be conceived as two circularly polarized rays with equal amplitudes and opposite senses of rotation. The dextrorotatory and laevorotatory polarized rays travel in an optically active medium at different velocities (they have different refractive indices). Phase shift thus takes place and after rejoining on exit from the optically active medium the ray oscillates in a plane deflected from the original plane.

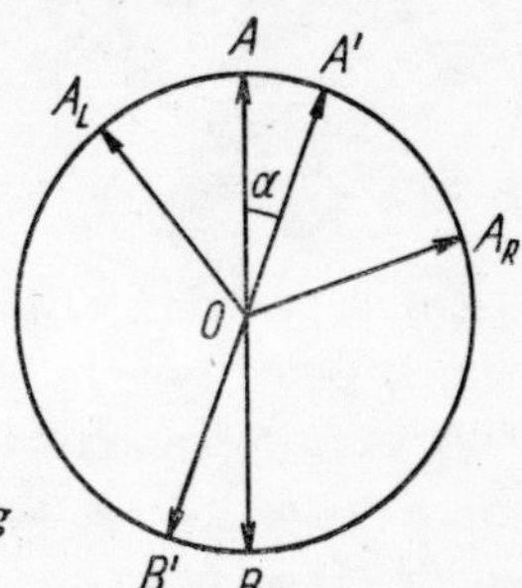

Fig. 5. Rotation of plane of the polarized light according to Fresnel.

The optical activity of organic compounds requires chiral structure of the molecule together with the chiral distribution of valence electrons which are manifested in the individual chromophores of the electronic spectra. From this aspect, we may imagine the molecule as a system of electronic oscillators which all add to the total activity by their contribution, even if small. On such a basis, the **principle of superposition** may then be defined as follows. The optical rotation of a compound is the sum of the rotation contributions of the individual absorption bands. The contribution of each chromophore band is given both by its intensity and by its anisotropy, which is the result of the asymmetric coupling of the motion of electrons located at the individual bonds. In connection with optical activity, each group of atoms may bring about two effects: either characteristic electronic transitions occur in it and may become optically anisotropic as a result of the effect of the remaining oscillators, or the vicinal effect will influence the chromophores in the vicinity of the group. Especially powerful is the effect of relatively long-wave electronic transitions with low intensity in the ultraviolet spectrum, as for example the n-π*transition of the ketonic carbonyl group. Besides optically active chromophores, the anisotropy of which is induced by the effect of its chiral surroundings, we of course also encounter internally (inherently) chiral chromophores, in which the valence electrons occupy an enantiomeric configuration even in a non-chiral environment, *e.g.* biphenyl compounds (*II*).

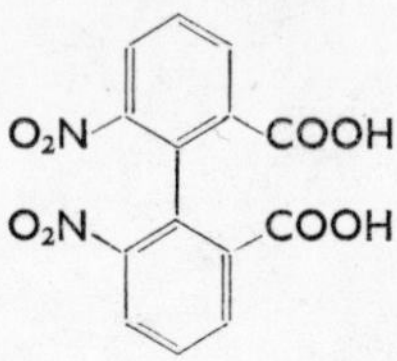

O_2N COOH
O_2N COOH

II

The dependence of optical rotation on the wavelength of polarized light is called **rotatory dispersion** (7, 8). It expresses the change of the differences in absorption of the dextrorotatory and laevorotatory circularly polarized beams (see Fig. 5) accompanying the change of the wavelength. The curve (the so-called normal, plain one) illustrating this dependence in the region far removed from the absorption bands of the molecule is monotonous, with no peaks, minimum values or points of inflection. However, the absolute rotation value increases in the direction of lower wavelengths. This effect is already clearly discernible in the range of visible light. The shape of the curve in this region is given by the **Drude equation**:

$$[\alpha]\ T\ ^\circ\mathrm{C} = \Sigma\ K_i(\lambda^2 - \lambda_i^2),$$

where K_i and λ_i are constants of the substance and λ is the wavelength of the light used. The equation expresses the sum of the contributions of all absorption bands. The shape of the curve begins to be interesting in the wavelength region corresponding to the absorption band of the optically active chromophore. The original monotonous shape changes rapidly, attains two extrema and forms the so-called curve of anomalous dispersion. In this region of the spectrum we may also observe the different absorption of dextrorotatory and laevorotatory polarized beams, the so-called **circular dichroism**, leading to the transformation of the original circularly polarized light into light that is elliptically polarized.

The curve illustrating the change of the differences of both absorptions (the so-called **ellipticity**) has the shape of an isolated absorption band and attains non-zero values only in that part of the spectrum in which the optically active chromophore absorbs. Both anomalies together are called the **Cotton effect**. The point of inflection of the Cotton effect curve on the dispersion curve is located in the vicinity of the maximum of the curve of circular dichroism and, as a rule, in the vicinity of the maximum of the corresponding band in the electronic spectrum (see Fig. 6)*. Both properties, optical rotatory dispersion as well as circular dichroism, are a manifestation of the same fundamental structural feature, namely of the chiral distribution of electrons. We may say that rotation is caused

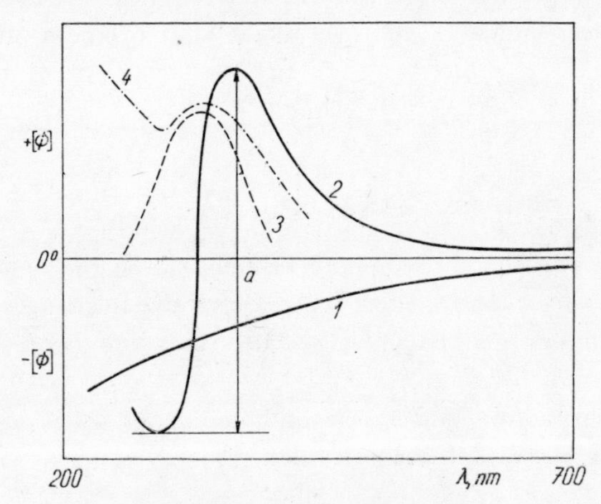

Fig. 6. Optical rotatory dispersion.

1 – plain curve, 2 – curve of anomalous dispersion with positive Cotton effect, 3 – curve of positive circular dichroism, 4 – absorption in the ultra-violet region, a – amplitude of the Cotton effect.

---

* We designate as positive those rotatory dispersion curves which tend to positive values when proceeding in the direction of shorter wavelengths. We speak of a positive Cotton effect when the curve first of all forms a maximum and then a minimum. The rotation value for comparison of the results of measuring various compounds is expressed as **molecular rotation** $[M]$ or $[\Phi] = [\alpha]$. mol. weight/100. The size of circular dichroism is expressed either as **molecular ellipticity** $[\Theta]$, or as **dichroic absorption** $\Delta\varepsilon$. Both values are interconnected by means of the simple relationship $[\Theta] = 3300 . \Delta\varepsilon$.

by the motion of electrons and dichroism by the absorption of light energy, which is itself caused by the former. From the experimental point of view, both manifestations are absolutely independent and not interchangeable; their practical applications of course overlap very considerably (9).

Empirical Relationships between Rotation and Structure

On the basis of previous considerations we may formulate some of the empirical rules.

a) In a series of analogous compounds, the absolute value of rotation is generally larger, the closer the respective optically active bands are to the visible region (the influence of the shift of the Cotton effect):

$$CH_3-\underset{\displaystyle X}{\underset{|}{CH}}-CON(CH_3)_2$$

$[\Phi]$ for X = Cl −82°; Br −152°; N_3 −259°

b) Changes in the chromophore bringing about the optically active band have a very intensive effect, whereas that due to changes in adjacent atoms is much less marked and its size decreases with increasing distance from the asymmetric atom.

Thus the rotations of members of homologous series are relatively constant (the so-called **Tschugaeff rule**):

$$CH_3-\underset{\displaystyle OH}{\underset{|}{CH}}-R$$

R = C_2H_5	n-C_3H_7	n-C_4H_9	n-C_5H_{11}	n-C_6H_{13}	
$[\Phi]_D^{20}$ 10.3°	12.1°	11.8°	12.0°	12.7°	
	n-C_7H_{15}	n-C_8H_{17}	n-C_9H_{19}	n-$C_{10}H_{21}$	n-$C_{11}H_{23}$
	12.9°	13.7°	14.0°	14.5°	14.4°

c) On the other hand, even a small chemical change in the substituent directly adjacent to the asymmetric carbon atom affects the rotation characteristically according to the **rule of shift**: analogous compounds with the same configuration change their rotation in the same direction

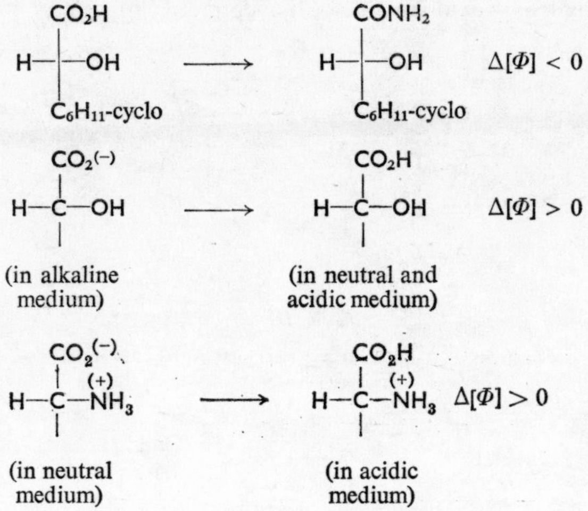

$$\begin{array}{c} CO_2H \\ H-\!\!|\!\!-OH \\ C_6H_{11}\text{-cyclo} \end{array} \longrightarrow \begin{array}{c} CONH_2 \\ H-\!\!|\!\!-OH \\ C_6H_{11}\text{-cyclo} \end{array} \qquad \Delta[\Phi] < 0$$

$$\begin{array}{c} CO_2^{(-)} \\ | \\ H-C-OH \\ | \end{array} \longrightarrow \begin{array}{c} CO_2H \\ | \\ H-C-OH \\ | \end{array} \qquad \Delta[\Phi] > 0$$

(in alkaline medium) (in neutral and acidic medium)

$$\begin{array}{c} CO_2^{(-)} \\ | \quad (+) \\ H-C-NH_3 \\ | \end{array} \longrightarrow \begin{array}{c} CO_2H \\ | \quad (+) \\ H-C-NH_3 \\ | \end{array} \qquad \Delta[\Phi] > 0$$

(in neutral medium) (in acidic medium)

Diagram B 8

if the corresponding substituents are changed in the same way. An important representation of the specific conclusions following from the rule of shift is the **Hudson amide rule** (see Diagram B 8): all α-hydroxy acids the rotation of which is shifted to the right by their transformation into amides have configuration *III*. The transformation of the carboxylic

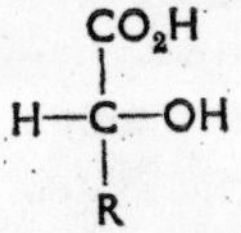

$$\begin{array}{c} CO_2H \\ | \\ H-C-OH \\ | \\ R \end{array}$$

III

group into the carboxylate anion is also accompanied by a positive shift of rotation in the case of substances with the absolute configuration shown (see Diagram B 8) regardless of whether we compare the pair α-hydroxy acid/carboxylate ion or the α-amino acid salt with mineral acid/free amino acid in aqueous solution (the so-called **Lutz-Jirgensons rule**). The **lactone rule** gained importance first in the carbohydrate chemistry and more recently has been used for a number of other compounds: a lactone with the configuration of the lactone-forming hydroxyl as shown in Formula *IV* has a more positive rotation than the corresponding hydroxy acid and also the epimeric lactone of Formula *V* (see Table III).

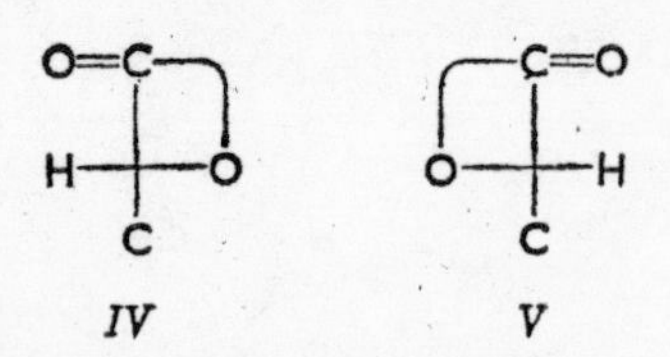

Table III

DIFFERENCES OF OPTICAL ROTATION OF EPIMERIC LACTONES

Type IV	*Type V*	$[\Phi]_D$ *IV* – $[\Phi]_D$ *V*
D-*allo*	D-*gulo*	+90°
D-*ribo*	L-*xylo*	+151°
D-*gluco*	D-*galacto*	+258°
D-*xylo*	L-*arabino*	+248°

d) A general extension of these rules is the **method of molecular rotation differences**, applied mainly to the investigation of steroid and terpenoid compounds. On the assumption that a saturated polycyclic hydrocarbon, built of strain-free six-membered and five-membered rings, has a negligible molecular rotation, and that the individual centres of the relatively rigid polycyclic system are mutually independent to a considerable extent, we may express the molecular rotation of a steroid derivative as the sum of rotation increments of its individual significant parts. The difference $\Delta[\Phi]$ between rotations of the investigated derivative and the simpler derivative, preferably the parent hydrocarbon, is characteristic of the given structure and configurations. With the help of the known values of rotation increments we may then decide as to the character (structure, position and configuration) of the substituent and, by comparing the $\Delta[\Phi]$ values, carry out the correlation of the two compounds, which are difficult to correlate chemically. As an example we may introduce the determination of the configuration of the carbon atoms $C_{(12)}$ and $C_{(13)}$ in abietic acid (*VI*) by comparison with

3,5-cholestadiene (*VII*); in both cases the rotation of the unsaturated compound is much more negative than that of the saturated compound

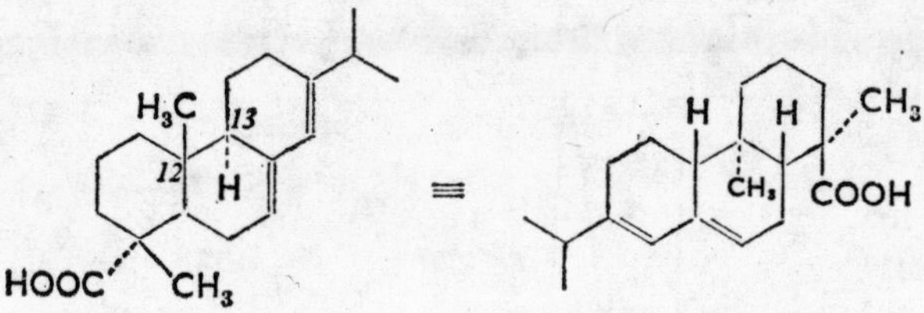

VI

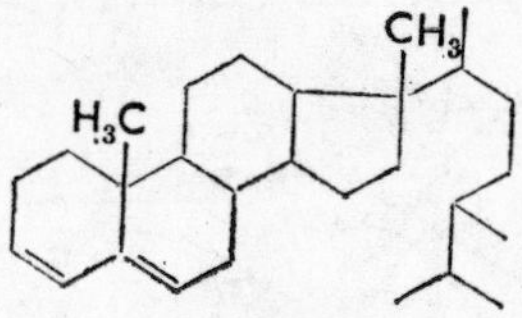

VII

$\Delta[\Phi](C{=}C{-}C{=}C)$
in compound *VI* = −335°
in compound *VII* = −549°

Application of Rotatory Dispersion

In the practical application of rotatory dispersion (7, 8) we utilize the considerable increase of the absolute value of rotation as well as the characteristic shape of the dispersion curves. The first factor enables us to work with very small samples (1 mg), to carry out a quantitative analysis of a mixture of optically active compounds and also to decide unequivocally whether a compound is optically active or not. The shape of the curves is useful for structural investigations comprising the determination of the position of the chromophore group, the spatial arrangement of the substituents in its vicinity as regards relative configuration as well as the details of conformation (see p. 93), and finally absolute configuration.

When determining the absolute configuration by comparing the rotatory dispersion curves of two compounds of the same structure in the critical region, we depend on the finding that substances of the same absolute configuration manifest the same gross orientation of the dispersion curve and especially of the Cotton effects. Steroid derivatives of known absolute configuration are frequently used as the basis for comparison. An illustrative example is that of yohimbine (*VIIIa*), pseudoyohimbine (*VIIIb*) and herbaceine (*IX*). The first two compounds have opposite absolute configuration on $C_{(3)}$ and show enantiomeric curves (Fig. 7).

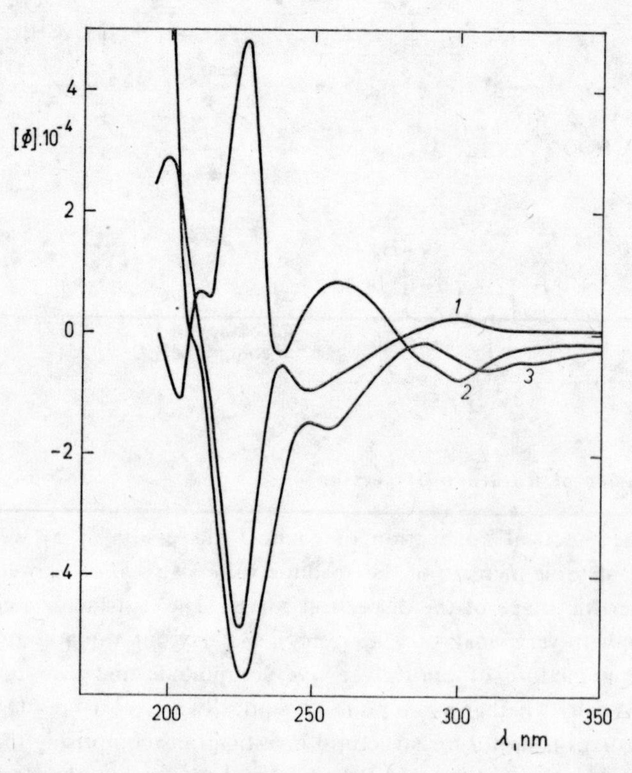

Fig. 7. Determination of the absolute configuration with the help of optical rotatory dispersion.

1 – Yohimbine, 2 – pseudoyohimbine, 3 – herbaceine. The curves 2 and 3 are approximate mirror images of each other and consequently the two substances are of enantiomeric character; the similar shape of the curves indicates a similar structure in the vicinity of the chromophores.

The ORD curve of herbaceine is very similar to that of yohimbine (*VIIIa*) but enantiomeric to that of pseudoyohimbine (*VIIIb*). The absolute configuration on $C_{(3)}$ of herbaceine (*IX*) in thus opposite to the known absolute configuration of pseudoyohimbine (*VIIIb*) and the same as for yohimbine (*VIIIa*):

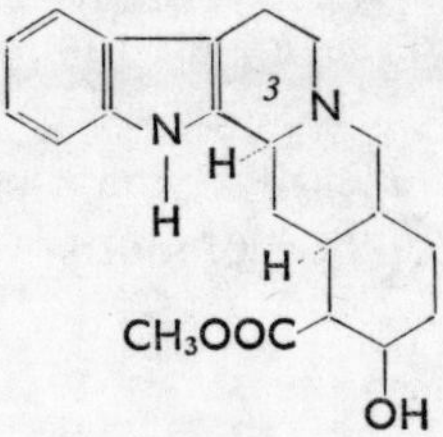

VIIIa 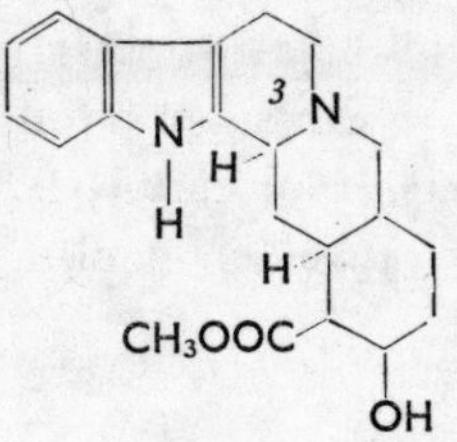 *VIIIb*

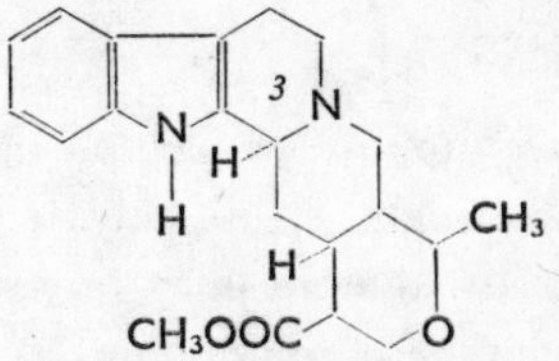

IX

Also interesting is the determination of the absolute configuration of lysergic acid (*X*) by comparing it with $\Delta^{1,10b}$-*N*-methylhexahydrobenzo[*f*]quinolin-2-ols (*XI*). Direct chemical correlation is in this case difficult to imagine.

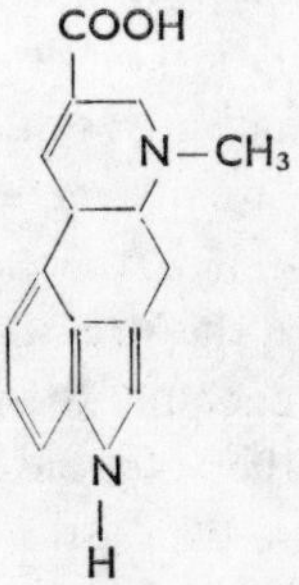

X 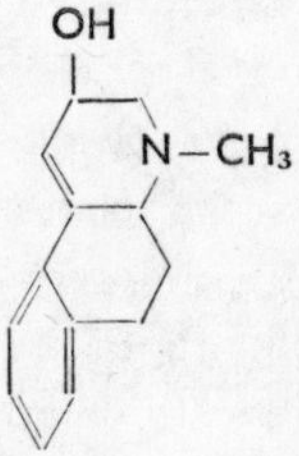 *XI*

Determination of absolute configuration with the help of plain dispersion curves lacking the Cotton effect is rather unreliable and therefore it is always advisable to introduce a new substituent into the investigated compound, which would incite this effect. As an example we may use L-α-amino acids and L-α-hydroxy acids, which themselves have only unexpressive dispersion curves. Certain of their sulphur-containing derivatives (dithiocarbamates (*XII*) and analogous xanthates (*XIII*)), however, show a marked positive Cotton effect. Of similar importance is the conversion of carboxylic acids branched at the α-position into thiourea derivatives or of olefins into addition products with osmium

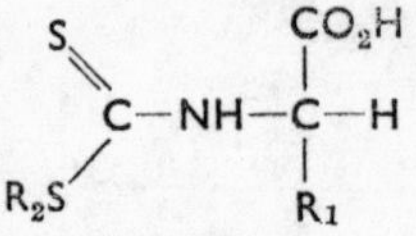

R_2S–C(=S)–NH–C(H)(CO_2H)(R_1)

XII

R_2S–C(=S)–O–C(H)(CO_2H)(R_1)

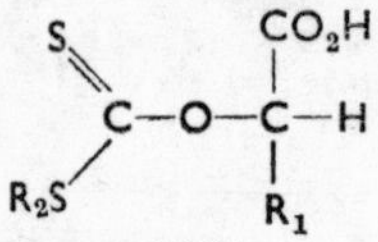

XIII

tetroxide. Very sensitive up-to-date equipment permits measurement even at very short wavelengths (routinely to 200 nm, in a nitrogen atmosphere to 185 nm), so that we may record even the Cotton effects corresponding to short-wave chromophores (*e.g.* the carboxyl group).

Octant Rule

The application of the previously described methods of determining absolute configuration, based on measurements of optical activity, requires the comparison of optical rotation changes of the investigated compound with rotation changes of several comparative substances of known absolute configuration. The basic condition for success is the choice of comparative substances sufficiently similar to the investigated compound. The use of unsatisfactory structural models may very easily lead to erroneous conclusions. It is therefore attractive to attempt to express the relationship between the structure of a substance and its rotation by a more exact method enabling the determination of the absolute configuration by measuring the rotation of a single compound or conversely to determine the sense of rotation and the sign of the Cotton effect, possibly even to semiquantitatively estimate their size, on the basis of the known stereochemical arrangement of a compound.

Systematic investigation of optical rotatory dispersion curves of ketones of various types led to the formulation of the so-called **octant**

rule, which currently represents the most successful attempt to comprehend the relationship between structure and rotation in a more exact manner. In its original form, it was meant for substituted cyclohexanones only. However, it proved that its validity could, under certain conditions, be extended to the nearest higher as well as lower homologous cyclanones (cyclopentanones and cycloheptanones). Atropisomeric biphenyl derivatives, 1,3-dienes and unsaturated aliphatic ketones were treated in the same manner (10). It appears that thus "extended" the octant rule will be quite generally valid. Of course, in order to understand the principle of the rule we will keep to the cyclohexanone compounds.

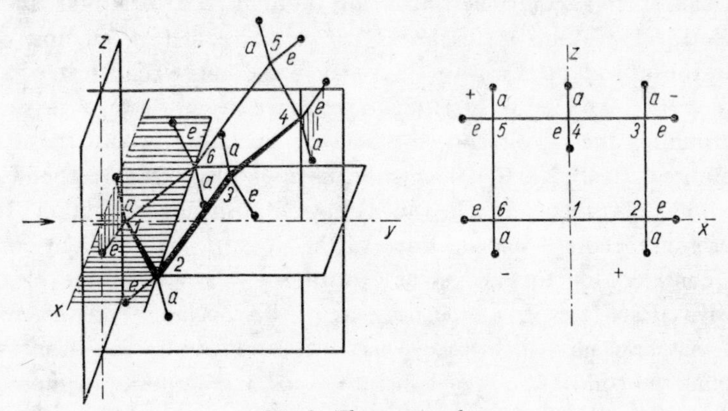

Fig. 8. The octant rule.

The cyclohexanone molecule may be orientated according to a system of three planes (see Fig. 8). The first one contains the carbonyl group and the two adjacent carbon atoms $C_{(2)}$ and $C_{(6)}$ (plane *xy*), the second one contains the oxygen atom, carbon $C_{(1)}$ and the opposite carbon $C_{(4)}$* (plane *yz*). Finally, the third plane intersects the double bond of the carbonyl group perpendicularly (plane *xz*). The planes divide the space in which the molecule is located into eight octants, the four front ones situated nearer to the observer when viewed in the direction of the O=C bond (see arrow in Fig. 8) and the four rear ones. The location of the substituents in the octants illustrates the mutual

---

* Or bisects the opposite side in the case of cyclopentanone or cycloheptanone.

position of the electrons contributing to the formation of an optically active band. By comparing the rotation values of a large number of cyclohexanone derivatives it becomes possible to allocate the sense of the contribution to the total course of the optical rotatory dispersion curve, shown by the molecular segments in the individual octants. Atoms located in the same octant show rotation contributions of the same sense; their contributions are added up. The atoms situated closest to the carbonyl group have the greatest effect. The atoms contained in one of the planes, *i.e.* the C=O group, carbons $C_{(2)}$, $C_{(4)}$ and $C_{(6)}$, as well as the equatorial substituents attached to $C_{(2)}$ and $C_{(6)}$ and both $C_{(4)}$ substituents, contribute practically nothing and therefore can be neglected. The atoms in the lower left rear octant (for example the axial substituent attached to $C_{(6)}$) and in the upper right rear octant (carbon $C_{(3)}$ and both substituents attached to it) have a negative contribution; the substituents in the lower right rear octant (the axial substituent attached to $C_{(2)}$) and in the upper left rear octant (carbon $C_{(5)}$ and both substituents attached to it) have a positive contribution. The other four octants situated closer to the observer only seldom come into consideration for analysis (*e.g.* with 1-oxosteroids). None of the decisive atoms are, as a rule, located in these octants and therefore they may be neglected. In case, where it is necessary to take them into account the contributions of their atoms are just opposite to those in the four rear octants (for example a group in the lower left front octant has a positive contribution). By evaluating the contributions of the individual atoms not contained in the planes (which partially eliminate one another), we establish the sign of the resulting contribution, which simultaneously provides the sign of the Cotton effect of the investigated substance. Conversely, if we know the sign of the Cotton effect from experimental measurement, we are, on the basis of the octant rule, able to judge as to the probable spatial distribution of atoms in the studied molecule, for example as regards its conformational arrangement.

The method of application may be demonstrated on the enantiomers of *trans*-hexahydroindanone (*XIV*), its homologue (*XV*) with an angular methyl group and similar compounds with two six-membered rings. The dextrorotatory enantiomer of *trans*-hexahydroindanone (corresponding to Formula *XIV*) shows a positive Cotton effect in agreement

XIV, R = H
XV, R = CH_3

with the requirements of the octant rule. The same is true of the methyl derivative (*XV*), where the methyl group is contained in one of the planes and thus cannot influence dispersion. The situation is different for *trans*-1-decalones. In the case of the parent ketone (*XVI*) we encounter a negative Cotton effect because the atoms $C_{(5)}$ and

XVI

$C_{(6)}$, which are the only ones not contained in some of the planes (as for instance $C_{(8)}$ and $C_{(7)}$) or without having a counterpart (like the pairs $C_{(3)}$ and $C_{(10)}$), are situated in an octant in which they have a negative contribution. However, in the case of the methyl ketone (*XVII*), the

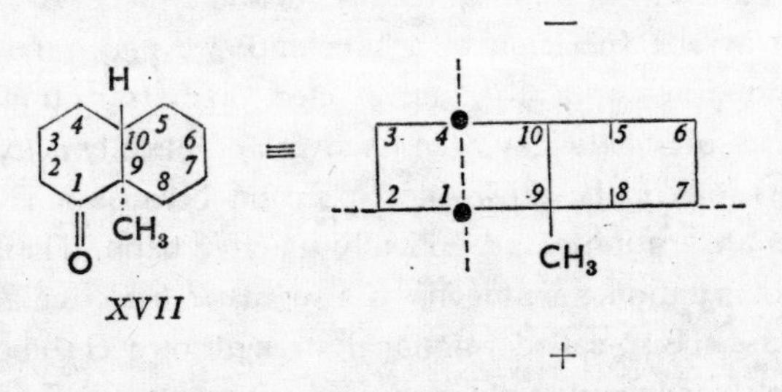

XVII

methyl group belonging to a positive contribution octant exerts an influence. As the methyl group is closer to the carbonyl group its influence prevails and the Cotton effect is positive, in agreement with experimental results. Analogous examples could be shown from the field of polycyclic and heterocyclic compounds.

Comparison of the Applicability of Optical Rotatory Dispersion and Circular Dichroism

As has already been said, the practical applications of both the above methods overlap considerably. It will therefore be expedient to summarize the advantages or disadvantages of both. The main advantage of optical rotatory dispersion is in its being measurable in the whole range of wavelengths (700 – 180 nm). It can therefore also be used for the investigation of substances with an inaccessible Cotton effect region (either because the respective substance absorbs too intensively or because the Cotton effect appears at shorter wavelengths than can be measured by the apparatus). The plain curves obtained may, after mathematical treatment, provide information on the spatial arrangement of polypeptides, for example. Anomalous dispersion curves are, furthermore, characteristically differentiated. The visual comparison of two or several curves, which is very frequent in organic chemistry, is then easier than the analogous comparison of circular dichroism curves. Finally, the apparatus for measuring rotatory dispersion is simpler and therefore more easily accessible.

The advantages of circular dichroism will appear especially in the course of a more exact investigation of rotation phenomena. The individual Cotton effects, which very often overlap on the dispersion curve, thus making the determination of their number and even of their sense sometimes difficult, are more readily distinguished. By comparison with the ultraviolet spectrum of a substance we may safely determine the corresponding optically active electronic transitions. Circular dichroism measurements may even uncover an optically active transition which is invisible in the ultraviolet spectrum because it is covered by the intensive absorption of an optically inactive band. The quantitative evaluation of rotation phenomena is also easier for circular dichroism curves because the so-called **rotational strength** of a chromophore may be calculated from the directly measured values. Finally, in the case of circular dichroism the measurements are more easily conducted at different temperatures (even considerably below −100 °C), which is important for conformational studies.

Other Physical Methods Suitable for Determination of Configuration

X-ray Studies

Under favourable conditions, X-rays permit the exact localization of the individual atoms in the molecule of a crystalline substance. In the case of organic crystalline substances we may thus determine not only the constitution but also the relative arrangement of the substituents attached to the individual asymmetric atoms, as well as the details of the conformational arrangement. Nowadays, as efficient computers are available, thedetermination of the structure of even very complex molecules, such as vitamin B_{12} or proteins (*e.g.* haemoglobin), is becoming reality. In order to determine absolute configuration, we may use the anomalous phase difference occurring with the diffraction of X-radiation, according to the procedure worked out by Bijvoet (1). The procedure made possible the safe determination of the absolute configuration of sodium rubidium tartrate, thus also verifying the fact that D(+)-glyceraldehyde, chosen decades ago as a standard for a correlation series, really has the absolute configuration assigned to it according to the Fischer convention. For the first time, the way was thus opened for an exact interpretation of the optical activity of organic compounds. Another advantage of this type of analysis is the possibility of determining absolute configuration even for other asymmetric atoms than carbon atoms, *e.g.* for the central atom of cobaltic complexes or for the sulphur atom in optically active sulphoxides. The absolute configuration of a large number of substances of various structural types has lately been determined by this method: several tens of compounds (11) in the field of alkaloids alone.

Nuclear Magnetic Resonance Spectroscopy

From the size of the coupling constants of two protons attached to two neighbouring carbon atoms we may determine the corresponding torsion angle (see p. 96) of the bonds attaching the protons. As pairs of diastereoisomeric substances usually differ as regards the value of this torsion angle in preferred conformations, it is possible to determine the relative configuration of the diastereoisomers from the coupling constant values, as has been carried out for example with the derivatives of ephedrine and pseudoephedrine or with the pair of amino acids isoleucine and alloisoleucine. Particulars may be found in special literature on nuclear magnetic resonance spectroscopy (12).

Quasi-Racemates

When discussing the nature of racemic substances, we illustrated the individual possibilities with the help of a diagram of the melting points of mixtures of optical enantiomers (see Fig. 2). A similar diagram, however, is shown by certain pairs of substances which, although not enantiomers, are structurally similar and with a mutually enantiomeric spatial arrangement. Here also we may assume the formation of a eutectic mixture as well as the formation of a solid solution without an expressive extremum on the curve, and finally an analogy with the formation of a racemic compound with the two enantiomeric components in ratio 1 : 1. The last situation occurs in cases when the two substances are very similar as regards structure, so that they may isomorphically represent real enantiomers in a crystalline lattice. Fredga (13) therefore utilized the so-called **quasi-racemate formation procedure** for the determination of the absolute configuration of compounds which cannot be mutually correlated by chemical means, especially for compounds with a carbon chain branched just at the asymmetric carbon atom. A significant example was the determination of the absolute configuration of (+)-methylsuccinic acid by comparison with (−)-mercaptosuccinic acid and (−)-chlorosuccinic acid, because the absolute configurations of most natural terpenes could be related to methylsuccinic acid as the standard. An instructive example is also given by benzylsuccinic acid (*XVIII*) and α-thiophenylmethylsuccinic acid (*XIX*).

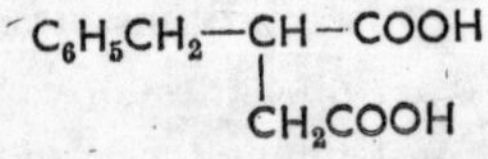

$C_6H_5CH_2-CH(CH_2COOH)-COOH$

XVIII

$C_4H_3S-CH_2-CH(CH_2COOH)-COOH$ (thiophene ring)

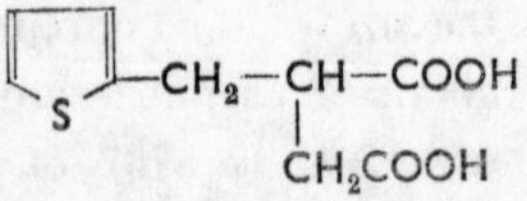

XIX

The absolute configuration of the first acid is known from chemical correlation. The diagram of the melting points of its two enantiomers is a typical example of the formation of a racemic compound (see Fig. 9a).

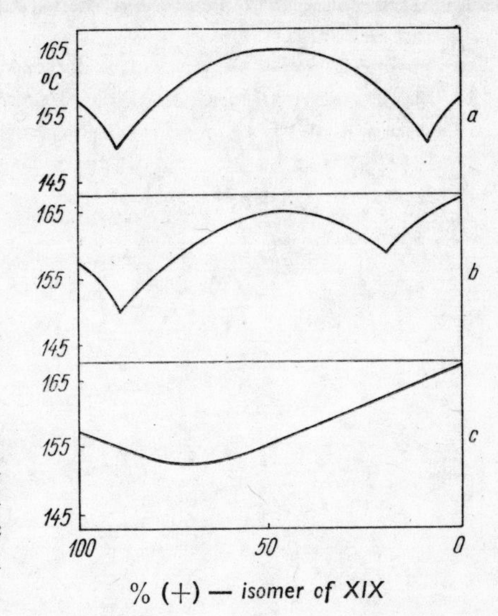

Fig. 9. Diagrams of mixed melting points of (+)-isomer of α-thiophenylmethylsuccinic acid (XIX).
a) with (−)-isomer of acid *XIX*, formation of racemic compound; b) with (−)-isomer of benzylsuccinic acid *(XVIII)*, formation of quasi-racemic compound; c) with (+)-isomer of acid *XVIII*, formation of eutectic mixture.

The diagram of the melting points of the mixture of the dextrorotatory enantiomer of acid *XVIII* and of the laevorotatory enantiomer of acid *XIX* (Fig. 9b) shows the formation of a quasi-racemate. The two substances thus have opposite configurations at the asymmetric carbon atom.

Literature

1. Bijvoet J. M.: Endeavour *14*, 71 (1955).
2. Bláha K., Kovář J.: Chem. Listy *56*, 129 (1962).
3. Tichý M.: *Advances in Organic Chemistry*, Vol. 5, p. 115. Interscience, New York, 1965.
4. Prelog V., Höfliger O.: Helv. Chim. Acta *33*, 2021 (1950).
5. Klyne W. in: *Determination of Organic Structures by Physical Methods*, Vol. 1, (E. A. Braude, F. C. Nachod, Eds), p. 73. Academic Press, New York, 1955.
6. Klyne W.: *Advances in Organic Chemistry*, Vol. 1, p. 239. Interscience, New York, 1960.

7. Djerassi C.: *Optical Rotatory Dispersion.* McGraw-Hill, New York, 1960.
8. Crabbé P.: *Optical Rotatory Dispersion and Circular Dichroism in Organic Chemistry.* Holden-Day, San Francisco, 1965.
9. Velluz L., Legrand M., Grosjean M.: *Optical Circular Dichroism: Principles, Measurements and Applications.* Verlag Chemie and Academic Press, Weinheim and New York, 1965.
10. Mislow K.: Ann. N. Y. Acad. Sci. *93*, 457 (1962).
11. Macintyre W. M.: J. Chem. Educ. *41*, 529 (1964).
12. Wiberg K. B., Nist B. J.: *The Interpretation of NMR Spectra.* Benjamin, New York 1962.
13. Fredga A.: Tetrahedron *8*, 126 (1960).

C. CONFORMATIONAL ANALYSIS

The examples of isomerism introduced previously (except atropisomerism) concerned the mutual relationship of isomeric substances which differ from each other by the way in which the bonds between the individual atoms are arranged, *i.e.* by their **configuration**. Without breaking one or more of these bonds it is impossible to convert one isomer into the other one; each of them is an independent chemical individual with characteristic physical and chemical properties. At the same time we assume that free rotation around a single bond does not lead to the formation of further isomers.

On the other hand, the term **conformation*** covers different spatial arrangements of a certain array of atoms, following from free rotation around single bonds. The development of conformational analysis and its modern application were made possible particularly by the work of Hassel, Pitzer, Prelog and Barton (1, 2, 3).

The possibility of rotating two parts of a molecule around the single bond connecting them is influenced by the attraction or repulsion of all atoms not immediately bound to one another. Classical stereochemistry did not, as a rule, take into account these nonbonded interactions. In fact, some compounds are able to change their conformation only with difficulty and in such cases the problem of conformation merges with the problem of configuration (for example polycyclic systems).

An illustrative example is the case of *o, o′*-disubstituted biphenyl derivatives, which are atropisomeric in extreme cases. The transition to very mobile aliphatic molecules is gradual. Whereas a configurational

---

* In the course of analysis of non-cyclic systems we also encounter the term "rotational isomerism". It is used, however, more frequently by physical chemists than by organic chemists.

formula expresses unequivocally the structure of a molecule, no conformational formula is the only possible representation of the shape of the molecule but constitutes just a picture of one of the possible states occupied, under the given conditions, by a larger or smaller number of molecules of the substance. All these states may be mutually interconverted without breaking any of the covalent bonds and represent a single chemical individual. Even though conformational isomers (conformers, rotamers) are in constant equilibrium and all of them may be present in a sample of a certain compound, one of them is, as a rule, of lowest energy and determines the properties of the compound to a great extent.

Conformation of Aliphatic Compounds

The basic problem may best be explained by the simple example of the n-butane molecule. In the course of the rotation of both halves around the $C_{(2)}$—$C_{(3)}$ bond, the molecule passes through an infinite number

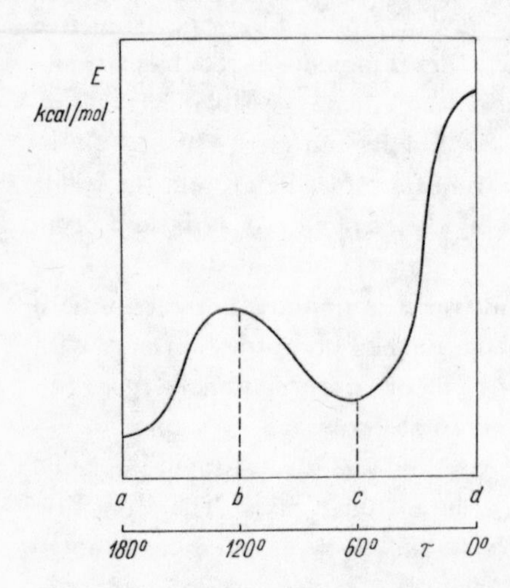

Fig. 10. Change of non-bonded interatomic energies during rotation of halves of n-butane molecule around $C_{(2)}-C_{(3)}$ *bond (schematic).*
On the *x*-axis – torsion angle $\tau°$; on the *y*-axis – energy change.

of conformations which differ as regards the relative position of the atoms attached to $C_{(2)}$ and $C_{(3)}$. We will consider those which are, in a certain respect, different from the rest of the others. The potential energy of a molecule is affected by the contributions resulting from the mutual attraction and repulsion of atoms not directly linked by bonds. During rotation this contribution changes with the change of the distances between the individual atoms. The extrema of the curve giving us the change of the interatomic non-bonded energies in the course of rotation* (see Fig. 10) specify the four conformations expressed by the diagramatic formulas *Ia—Id*. In the situation illustrated by point *a*, all the $C_{(2)}$ substituents are orientated against the gaps between the

* The transition from point *a* to point *d* represents a rotation of 180°. Rotation of another 180° could be observed on the second half of the curve, which would be a mirror image of the part shown. Considering the structural identity of the $C_{(2)}$ and $C_{(3)}$ substituents, we would thus arrive at the same conformational relationships as shown by formulas *Ia – Id*. Therefore they need not be discussed separately.

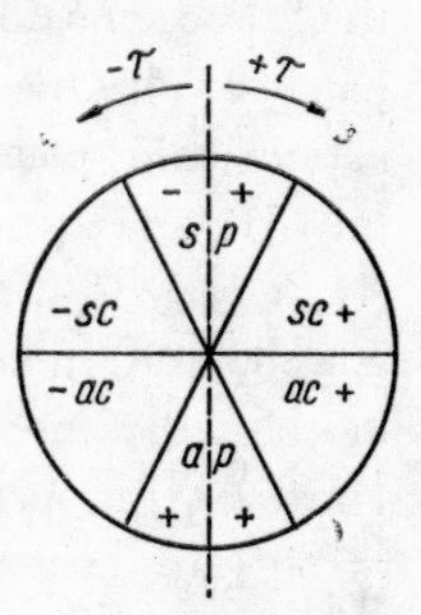

Fig. 11. Diagram for deriving conformational prefixes.

Older terminology used a number of not exactly defined notations for the designation of the individual conformations. Some of them are still being used: *staggered* for conformation *Ia*, *eclipsed* for *Id* and *skew* or *gauche* for *Ic*. The new terminology by Klyne and Prelog, which is more precise, is used throughout this book. The terminology designates the relative position of the two most significant substituents attached to neighbouring atoms by notations or symbols derived from Diagram 11. The *sp* position is called *syn*-periplanar, *sc syn*-clinal, *ac anti*-clinal and *ap anti*-periplanar. The importance of the substituents is assessed according to the Cahn-Ingold-Prelog system for each of the two neighbouring atoms independently, with two supplementary rules.

a) When two substituents attached to the same atom are identical, the conformation is determined with reference to the third one, without considering its seniority.

b) When all the substituents attached to one of the atoms are the same, we select the notation corresponding to the smallest torsion angle.

These notations are of course not sufficient for an exact description of the conformation and it is necessary to state the value of the torsion angle τ and its sign.

substituents attached to $C_{(3)}$ and *vice versa* so that there is a maximum distance and minimum interaction between all substituents. Moreover, the two most bulky substituents – the methyl groups – are turned away from each other (torsion angle τ*, 180°) and cannot interact at

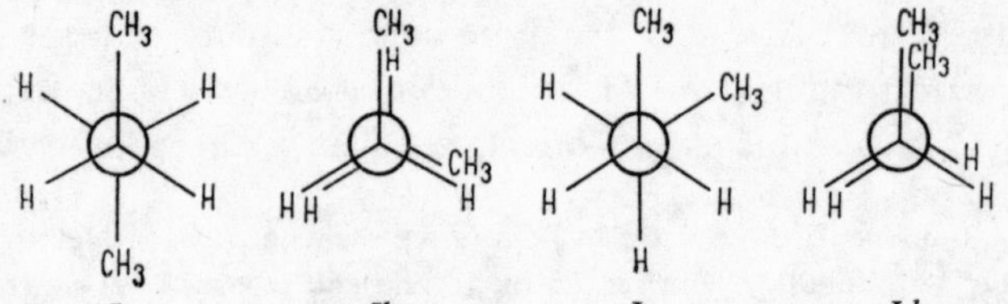

Ia *Ib* *Ic* *Id*

all. By a rotation of 60°, the bonds attaching the individual substituents to $C_{(2)}$ become coplanar with the bonds issuing from $C_{(3)}$; the respective pairs of substituents thus get closer to each other and repulsive interactions take place between them, one between the hydrogen atoms (H—H) and two between the hydrogen atom and the methyl group (H—CH_3). Consequently, the potential energy of the molecule rises considerably (maximum *b* in Fig. 10). With a further rotation of 60°, most of the interaction due to the coplanarity of bonds disappears. However, the methyl groups, attached by bonds with a torsion angle of 60°, are evidently near enough to each other to display a definite interaction (minimum *c*). Finally, by a total rotation of 180°, we convert conformation *Ia* into *Id*, where there again are three pairs of coplanar bonds and moreover the methyl groups are situated directly opposite each other (torsion angle 0°) so that they become the cause of a marked non-bonded interaction and of an increase of potential energy (maximum *d*). Only the minima of the curve can correspond to independently realizable conformational states which are the more stable the lower the potential energy contribution and especially the higher the energy barriers separating them from other minimum potentials. From the shape of the curve in the vicinity of a minimum it is evident that a significant energy contribution will result only by rotation through a relatively

* The torsion angle is an angle formed by two bonds issuing from two atoms linked by a chemical bond, projected on to a plane perpendicular to this bond.

large angle; a deflection of a few degrees only has no substantial effect and, without doubt, similar deformed partial conformations appear in the molecules of most compounds. The conformational situations expressed by the maxima of the curve have, however, to be taken into account in considering partial conformations in more complex compounds, namely cyclic ones, and in calculating the energy barriers limiting rotation.

An analogous analysis may be carried out for all ethane derivatives with the assumption that the polar effect of the substituents will not act significantly against their steric action and that none of the possible conformations is fixed by means of a hydrogen bond or another interaction. It will be expedient to consider from this point of view the case of aliphatic diasteroisomeric substances (see p. 23), differing in configuration at the two neighbouring carbon atoms. The most stable conformations of substances expressed in the Fischer projection by formulas *IIa* and *IIb* are given by the diagramatic formulas *IIIa* and *IIIb*. Considering that the repulsive forces are in principle larger between identical or similar substituents (M_1-M_2 and $S-S$) than between different substituents (M_1-S, M_2-S), we expect that the first spatial arrangement will be more stable than the second one, *i.e.* an *erythro*-compound will be more stable than a *threo*-compound and, for the case where $M_1 = M_2$, the non-resolvable *meso*-form will be the more stable one.

The conformational analysis of aliphatic compounds will also be applied to the study of asymmetric reactions (see p. 207).

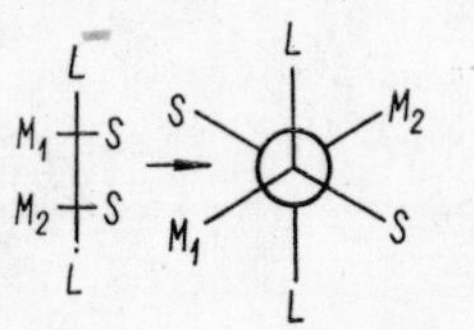

IIa *IIIa* 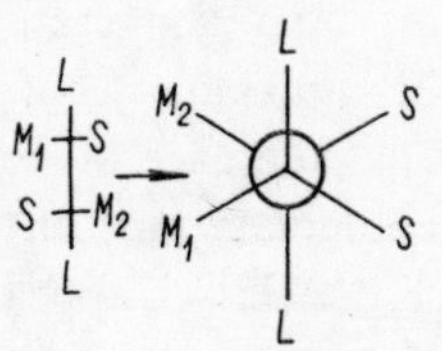 *IIb* *IIIb*

Conformation of Cyclic Compounds

The most fundamental conformational considerations, qualitative or quantitative, necessarily must be based on consideration of simple aliphatic molecules. However, for experimental investigations and, to a certain extent, even for theoretical interpretation, alicyclic compounds appear to be much more suitable, especially cyclohexane and its derivatives. The changes of potential energy caused by changing the conformational arrangements at the individual bonds may be derived from the analogous changes accompanying the change of conformation of a four-carbon segment of a polymethylene chain.

Already at the end of the last century, Sachs had established that it is possible to conceive of two forms of cyclohexane in which tetravalent carbon atoms maintain normal bond angles if we admit that all six atoms are not contained in one plane. The cyclic systems thus formed are completely without strain due to the deformation of bond angles and the compounds approach aliphatic compounds as regards their stability. The two stated forms are today called the **chair form** (*IV*) and the **boat form** (*V*).

If the partial conformation analogous to formula *Id* (*syn*-periplanar) represents a state of maximum potential energy, then the boat form necessarily has a higher energy content as it includes two partial arrangements of this type. The *syn*-periplanar arrangement in the polymethylene chains is characterized by a potential energy increase of 2.7 kcal/mol. The difference in energy between the two conformations thus amounts to at least 5.4 kcal/mol and is further increased by the interaction of two hydrogen atoms in positions 1 and 4. The energy barrier separating both forms is over 10 kcal/mol high but not sufficient to enable the isolation of the individual conformers as chemical individuals at room temperature*; it permits a very rapid interconversion of the individual conformers at a speed of the order of 10^6 per second. The difference of potential energy is, however, sufficient to make the predominant part of the molecules (over 99 %) of cyclohexane and its simple derivatives appear in the lower-energy chair form, under normal conditions. The boat form may appear in greater concentration only at higher temperatures, when sufficient energy has been supplied to permit the molecule to convert into a higher-energy state. Qualitati-

* Recently individual conformers have been isolated at low temperature and their nuclear magnetic resonance spectra measured.

vely, this is true of many heterocyclic analogues of cyclohexane (piperidine, 1,3,5-triazacyclohexane, *etc.*) although, at this moment, we do not have enough data for a detailed analysis.

When imagining the cyclohexane molecule in a chair form we see that always trios of non-vicinal carbon atoms are contained in two parallel planes. The bonds coming from them are of two types. The first – **axial** – ones are perpendicular to the planes formed by the non-vicinal atoms (Formula *VI*), whereas the second – **equatorial** – ones (Formula *VII*) are only slightly deflected from these planes (by approximately 20°, alternately above and below the plane). Two chair conformations are possible for each monosubstituted derivative, one with the substituent in equatorial position, the other one with the substituent in axial position. Of course, in this case also the respective formulas do not represent different chemical individuals. Simple conformational analysis (of methylcyclohexane for example) indicates that the conformer with the substituent in equatorial position should be the lower-energy one.

The bond attaching the equatorial methyl group to the ring (Formula *VIII*) is *anti*-periplanar with respect to all non-vicinal C—C bonds and the introduction of a methyl group into this position will produce a relatively low increase of energy content. The analogous methylsubstitution in an axial position (Formula *IX*) leads, however, to the formation of larger non-bonded interactions because the torsion angles formed by the $C_{(2)}$—$C_{(3)}$ and $C_{(6)}$—$C_{(5)}$ bonds with the CH_3—$C_{(1)}$ bond are 60° (*syn*-clinal conformation *Ic*). Pitzer estimates the potential energy increase brought about by the transition of the polymethylene chain segment from the *anti*-periplanar conformation to the above conformation at 0.9 kcal/mol*. The difference in energy between the two possible conformations of methylcyclohexane is thus at least 1.8 kcal/mol and ensures that under normal conditions most of the molecules appear in the conformation with the equatorial substituent.

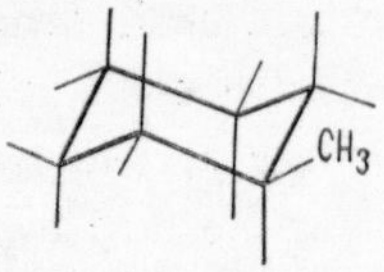

VIII 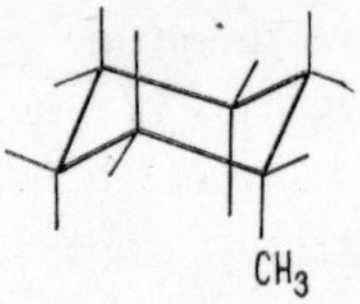 *IX*

Of course, the difference between the equatorial and axial substituents is not in their thermodynamic stability only but also in their steric accessibility. The substituent in an axial position is shielded, firstly by the cyclohexane ring and secondly by both substituents on the remaining axial positions which are on the same side of the molecule. The reactivity of the substituent is thus considerably influenced.

Similarly, with the chair and boat form of cyclohexane equilibrium is also established between both conformational isomers differing in the position of the substituent. The equilibrium depends both on the external conditions and on the character of the substituent. The influence of the effective steric volume of the substituent is usually decisive, but polar effects may, however, weaken the steric effect considerably. It is possible to calculate the **conformational equilibrium constant** K, the ratio of both isomers in the mixture, and finally the difference in free energies accompanying the transition of a substituent from the equatorial position to the axial position. The values for some of the usual substituents are shown in Table IV.

In the case of disubstituted cyclohexane derivatives, where geometrical isomerism leads to the formation of *cis*- and *trans*-isomers,

---

* The stated estimates of energy differences have to be considered rather as informative data and not as exact, experimentally verified parameters.

Table IV

DIFFERENCES OF FREE ENERGIES FOR SUBSTITUENTS IN AXIAL AND EQUATORIAL POSITIONS ON A CYCLOHEXANE RING AND PERCENTAGE OF EQUATORIAL ISOMER

Substituent	$-\Delta F$ (*kcal/mol*)	*Amount of equatorial substituent %*
CH_3	1.5–1.9	92–96
C_2H_5	2.1	97
Cl	0.3–0.5	62–70
1	0.4	65
OH	0.4–0.9	65–82
OCH_3	0.5–0.7	70–78
$OOCCH_3$	0.4–0.7	65–78
$COOC_2H_5$	1.1	85
COOH	1.6–1.7	93–95

the number of conformers is doubled. The table of conformational formulas of all disubstituted cyclohexanes (Table V) shows that only the *trans*-1,2-, *cis*-1,3- and *trans*-1,4-isomers can have both substituents in equatorial positions or both substituents in axial positions. The remaining three isomers are bound to have always one substituent in an axial position and the other one in an equatorial position. Conformational isomers again differ in their potential energy, which follows from the interactions of the atoms not linked to each other. In general, we may consider that the most stable isomer is the one in which both substituents are in equatorial positions; the isomer with one axial and one equatorial substituent is less stable and the least stable isomer is that with both substituents in axial positions. In the second case it is logical to assume the equatorial position will be occupied by the substituent with the larger effective steric volume, whilst the less bulky group is forced to stay in an axial position. The t-butyl group, which has the largest effective volume of all the substituents studied until now, remains permanently in an equatorial position on the chair conformation of the cyclohexane ring. It thus ensures that the position of the other substituents is also stabilized to a considerable extent because the transition of one chair conformation to another one is prevented. The fixation of conformation by introducing a t-butyl group into a molecule has been

Table V

CONFORMATIONAL ARRANGEMENT OF DISUBSTITUTED CYCLOHEXANES

(Substituent R_1 larger than R_2)

Isomer	*Conformation*
cis-1,2	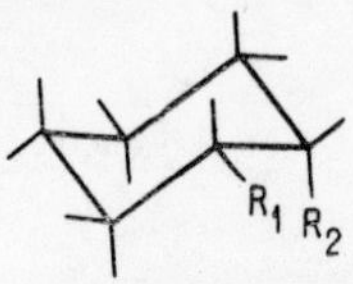 ea > 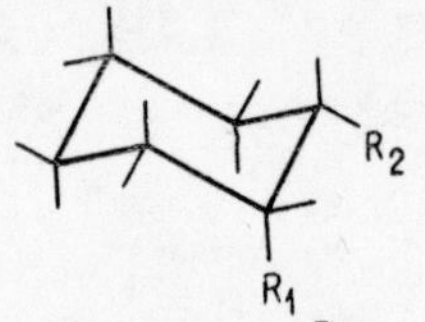 ae
trans-1,2	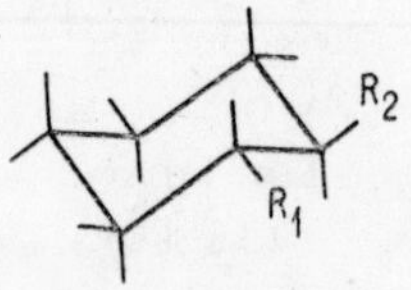 ee >> 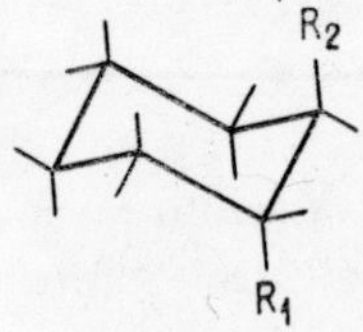 aa
cis-1,3	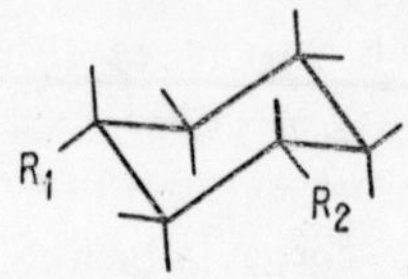 ee >> 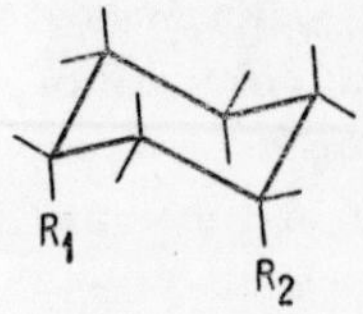 aa
trans-1,3	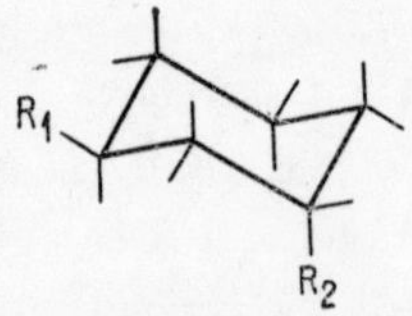 ea > 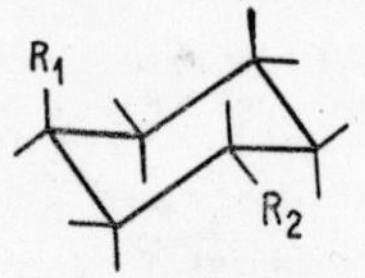 ae
cis-1,4	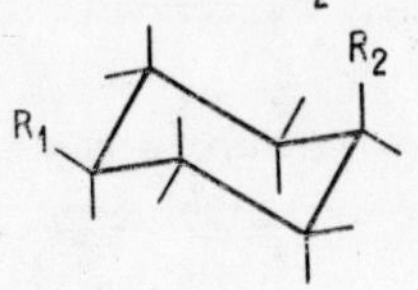 ea > 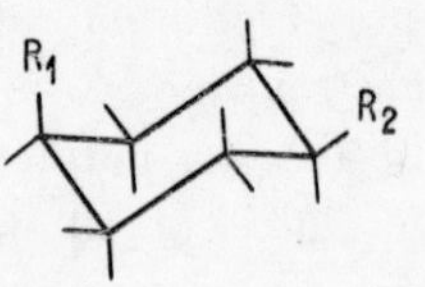 ae
trans-1,4	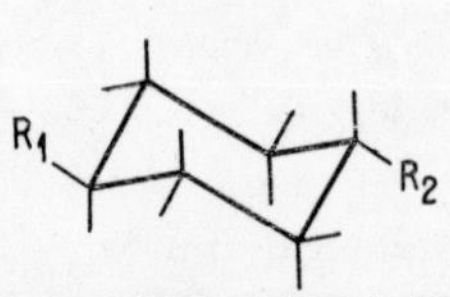 ee >> 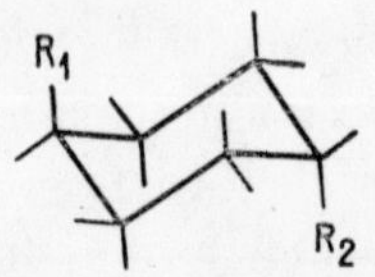 aa

utilized for the study of mechanism of reactions occuring on the cyclohexane ring and for the determination of the conformational stability of various substituents. Of interest is the comparison of stability of 1,2- and 1,3- diaxial isomers. In the 1,2-isomer, diequatorial substituents are relatively close to each other and if they are bulky enough they may hinder and repel each other partially into axial positions, whereas in the case of the 1,3-isomers the interaction of diequatorial substituents is not possible and it is the axial substituents that get very close to each other. According to Pitzer's calculations, *trans*-1,2-dimethylcyclohexane already contains at room temperature a small but demonstrable amount of the diaxial form. In the case of derivatives with strongly polar substituents this effect is still more marked; *e.g.* *trans*-1,2-dichloro- and 1,2-dibromocyclohexane contain 1/3 and 1/2 of the diaxial isomer respectively, as has been established by measurements of dipole moments and by electron diffraction*. In 1,3-isomers with both substituents in axial positions on the same side of the ring, they repel each other to such an extent that their interaction becomes the principal destabilizing factor. When they are forced to remain in axial position, a certain deformation of the bond angles and torsional angles takes place. Hexachlorocyclohexane isomers may serve as an example (see Table VI). Of the total number of eight possible isomers all four isomers in which 1,3-interaction cannot occur are known; of the other four only one isomer is known (γ-hexachlorocyclohexane (*X*)). Its existence is made

---

* The conformational analysis of disubstituted cyclohexane derivatives led to a revision of the so-called Auwers-Skita rule. According to its original version, the less stable *cis*-derivative formed during the catalytic hydrogenation of benzenoid raw materials has a higher energy content and differs from the *trans*-isomer in having a higher boiling point, higher density, higher refractive index, higher heat of combustion, but a lower molecular refraction. Accordingly, a number of cyclohexane derivatives were allotted their relative configuration. In the case of 1,3-disubstituted compounds, catalytic hydrogenation also yields a *cis*-isomer which is, however, thermodynamically more stable than the *trans*-isomer, on the basis of conformational analysis. The determination of relative configuration on the basis of the original version of the rule necessarily had to be incorrect in this case. Correct conclusions are drawn from the revised version of the rule according to which a lower boiling point, density, refractive index and heat of combustion but a higher molecular refraction value appertain to the thermodynamically more stable isomer.

Table VI

HEXACHLOROCYCLOHEXANE ISOMERS AND THEIR CONFORMATION

Designation	*Position of substituents*[a]	*Number of 1,3-interactions*
β	e e e e e e	0
ϑ	a e e e e e	0
α	a a e e e e	0
ε	a e e a e e	0
—	**a** e **a** e e e	1
γ	**a** a **a** e e e	1
—	a **a** e **a** e e	1
—	**a** e **a** e **a** e	3

[a] Bold-face axial atoms display mutual 1,3-interaction.

possible by the stabilizing effect of the axial substituent at position 2 (as will be shown). The isomer with three axial substituents on the same side is not capable of existence, like other compounds of this type.

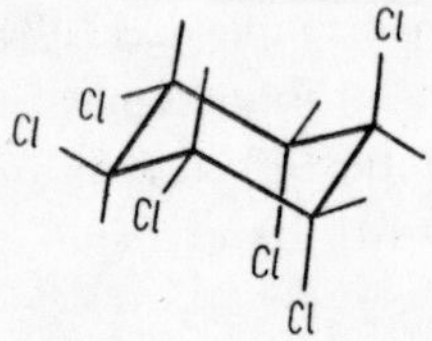

X

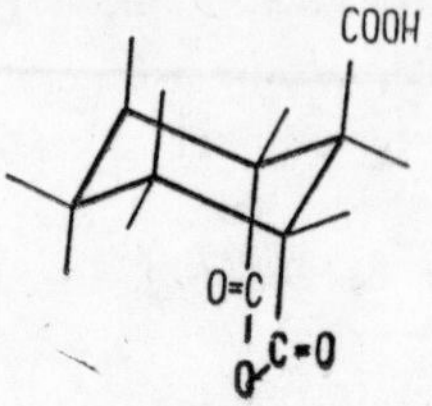

XI

The only exceptions are compounds in which the axial substituents are fixed by the formation of further rings (*e.g.* adamantane and its derivatives). It is of course very probable that even in these cases the bond angles and torsional angles of the individual parts of the molecule are changed. The cases of inverted order of stability are exceptional. Barton showed that a substituent at $C_{(2)}$ in compounds diaxially substituted in the 1,3-positions is more stable in the axial position than in the equatorial one. This anomaly can be illustrated by several examples of the stabilization of the axial position of the substituents at $C_{(1)}$ and $C_{(3)}$

either by bridging or by the effect of other substituents. Thus for example in the 1,3-anhydride of cyclohexane-1,2,3-tricarboxylic acid (*XI*) the free carboxyl at the carbon atom $C_{(2)}$ is more stable in the axial position; in the case of tropine (*XII*) it has also been established that the methyl group attached to the nitrogen atom occupies the axial position with respect to the piperidine ring. As has already been said,

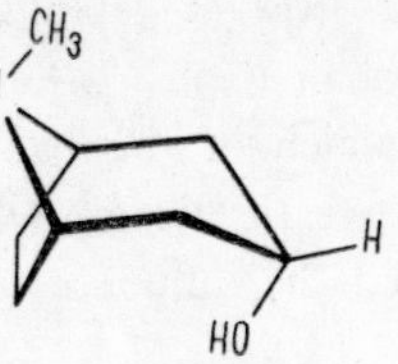

XII 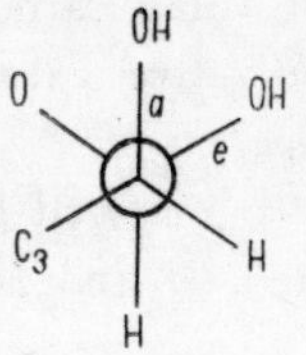 *XIII*

some substituents may, on the basis of their considerable polarity, stabilize a conformation other than the usual one. Whilst the methyl group in an equatorial position represents the lower-energy form of 2-methylcyclohexanone, in the case of α-halogenocyclohexanones the more stable conformation is with the substituent in the axial position, as has been established for example by spectral methods. The torsional angle contained by the carbonyl oxygen atom and the axial substituent is 105°, whereas the angle contained by the carbonyl oxygen atom and the equatorial substituent is only about 15° (see Diagram C 1). The repulsion between the C—halogen bond and the carbonyl group dipoles is in the second case much more powerful and overcomes the normal

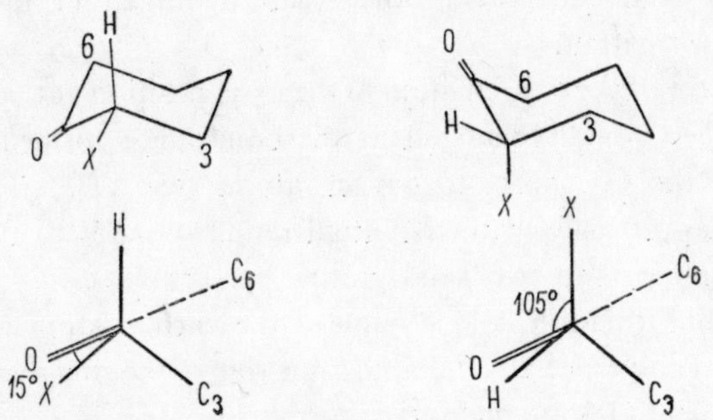

Diagram C 1

preference for a group to occupy the equatorial position. Similar, though less marked, repulsion occurs in the case of 2-alkylcyclohexanones with sterically demanding alkyl groups; of course in these cases it is a steric effect rather than a polar effect.

As an example of the application of conformational analysis to monocyclic compounds we may include the analysis of the stereochemical structure of some carbohydrate derivatives. Reeves established that pyranose derivatives exist predominantly in chair forms. In order to be able to judge, at least semiquantitatively, which of the two possible forms will be preferred by the given derivative (in Diagram C 2 they are designated *C1* and *1C*). Reeves evaluated the influence of various

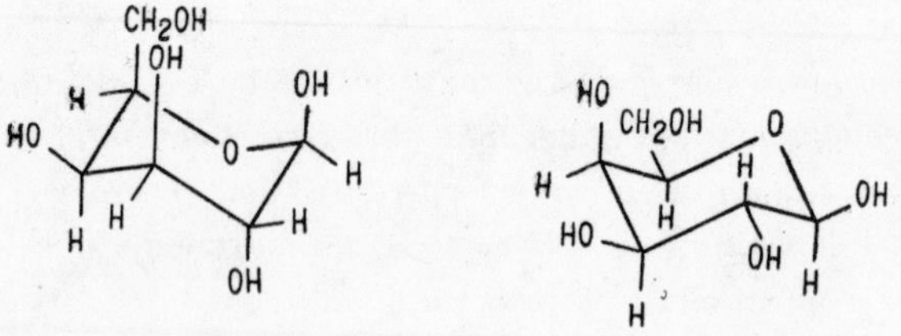

Diagram C 2

substituents on the stability of the basic ring and listed the following factors reducing the stability of pyranose rings.

A. Every axial substituent other than hydrogen brings instability into the conformation.

B. The effect of the oxygen atom in the axial position at $C_{(2)}$ is especially large if the $C_{(2)}$—O bond bisects the bond angle formed by the two bonds from the $C_{(1)}$ atom to oxygen atoms (see *XIII*). Three oxygen atoms thus approach and their mutual repulsion adds to the reduction of stability more than two axial groups by themselves.

C. The conformation is less stable if the carbon atom $C_{(6)}$ is in an axial position and especially if on the same side of the ring there is another axial substituent (1,3-interaction).

D. Stability is reduced to a still greater extent by carbon $C_{(6)}$ being present (as in case *C*) together with two further axial substituents on the same side of the ring.

As the survey of various aldopyranoses given in Table VII shows, each compound will always appear in the conformation in which it will be able to display a smaller number of the factors reducing stability. The results of the analysis are in full agreement with experimental results, *e.g.* the behaviour of the individual substances during the formation of complexes in cupric ammonium solutions. The isomer containing

Table VII

CONFORMATIONAL ANALYSIS OF SOME ALDOPYRANOSES

Compound	*Factors*[a] *reducing conformational stability*	
	C1	*1C*
α-D-Allose	A_{13}	B; A_{45}
β-D-Allose	A_3	C; A_{1245}
α-D-Altrose	B; A_3	C; A_{145}
β-D-Altrose	A_{123}	A_{45}
α-D-Galactose	A_{14}	C; B; A_{35}
β-D-Galactose	A_4	D; A_{1235}
α-D-Glucose	A_1	C; B; A_{345}
β-D-Glucose	none	D; A_{12345}
α-D-Gulose	A_{134}	B; A_5
β-D-Gulose	A_{34}	C; A_{125}
α-D-Idose	A_{1234}	A_5
β-D-Idose	B; A_{34}	C; A_{15}
α-D-Mannose	A_{12}	C; A_{345}
β-D-Mannose	B	D; A_{1345}
α-D-Talose	A_{124}	C; A_{35}
β-D-Talose	B; A_4	D; A_{135}
α-D-Arabinose	A_{123}	A_4
β-D-Arabinose	B; A_3	A_{14}
α-D-Lyxose	A_{12}	A_{34}
β-D-Lyxose	B	A_{134}
α-D-Ribose	A_{13}	B; A_4
β-D-Ribose	A_3	A_{124}
α-D-Xylose	A_1	B; A_{34}
β-D-Xylose	none	A_{1234}

[a] *A*, *B*, *C*, *D* are the factors reducing stability as stated in the text; the indexes at *A* designate the number of the carbon atom to which an axial hydroxyl or other substituent is attached.

fewer instability factors always predominates in the solution in the course of mutarotation. According to expectation, this is for example α-D-lyxose, β-D-galactose, β-D-glucose, α-D-mannose and α-D-talose.

Conformation of Polycyclic Compounds

The validity of our previous considerations has been extended to include also polycyclic systems incorporating one or several cyclohexane rings. In essence it is possible to say that the isomer which contains a larger number of C—C equatorial bonds at the common carbon atoms is the more stable one. As a rule, all the substituents on a polycyclic system are more stable in equatorial positions. By means of simple conformational analysis, using Pitzer's parameters for the individual conformational states of a four-carbon chain, we may estimate the difference in potential energies between the individual conformers. Isomeric decalins may serve to illustrate the procedure (Formulas *XIV* and *XV*). First of all we determine how often each partial conformation appears in each of the formulas. By means of an algebraic sum of the corresponding increments we then calculate the total difference of potential energy between the various isomers. Thus, in *trans*-decalin we may identify a total of 6 *anti*-periplanar arrangements with a torsional angle of 180° and 12 *syn*-clinal arrangements with a torsional angle of 60°. On the other hand, the *cis*-isomer comprises only 3 *anti*-periplanar arrangements but 15 *syn*-clinal ones. The *anti*-periplanar arrangement is the most favourable one and does not increase potential energy so that it need not be taken into account for calculation. There thus remains a difference of 3 *syn*-clinal partial arrangements between the *cis*-isomer (*XV*) and the *trans*-isomer (*XIV*), corresponding to a potential energy difference of 2.7 kcal/mol, which stabilizes the *trans*-isomer compared to the *cis*-isomer. Experimental determination of the intrinsic energy difference ΔU with the help of heats of combustion led to practically the same value. As it is possible to assume that the values of the $T\Delta S$ term in the equation $\Delta F = \Delta U - T\Delta S$ for the two isomers are near to each other, the preceding semiquantitative estimate was verified very well.

The introduction of an angular methyl group reduces the difference in energy between the *cis*- and *trans*-decalins. In 9-methyl-*trans*-decalin (*XVI*) the methyl group is necessarily in an axial position with regard to both rings, so that another four *syn*-clinal conformations are added to the list of partial conformations.

XIV, R = H
XVI, R = CH_3

XV, R = H
XVII, R = CH_3

On the other hand, in 9-methyl-*cis*-decalin (*XVII*) the methyl group is an axial substituent with respect to one ring only and contributes only two new C—C bond interactions in *syn*-clinal arrangement. The difference between *cis*- and *trans*-decalins, summed up above, is thus reduced to 0.9 kcal/mol. Certain heterocyclic compounds are even more stable with a *cis*-fusion of two six-membered rings, if an angular methyl group is attached (see Table VIIa).

Table VIIa

EFFECT OF ANGULAR SUBSTITUENT ON RELATIVE STABILITY OF ISOMERIC *cis*- AND *trans*- AMIDES

Substance	*Compositions of equilibrium*[a] *mixture, %*	
	cis	*trans*
$R^1 = R^2 = H$	32	68
$R^1 = H$; $R^2 = CH_3$	60	40
$R^1 = CH_3$; $R^2 = H$	71	29

[a] After being heated to 300 °C in the presence of palladium black.

The relative configuration is not fixed in the case of compounds in which the hetero atom is located at the junction of the rings. For aza-compounds, *cis*-fusion is less probable. As a rule, we encounter an analogue of the *trans*-decalin system. This was proven by means of X-ray diffraction for the simplest representative of the whole group – quinolizidine as well as for the complex polycyclic system of α-isosparteine (*XVIII*). On the other hand, in the isomeric sparteine (*XIX*) both quinolizidine parts cannot have the favoured *trans*-fusion but one half has to be analogous to *cis*-decalin.

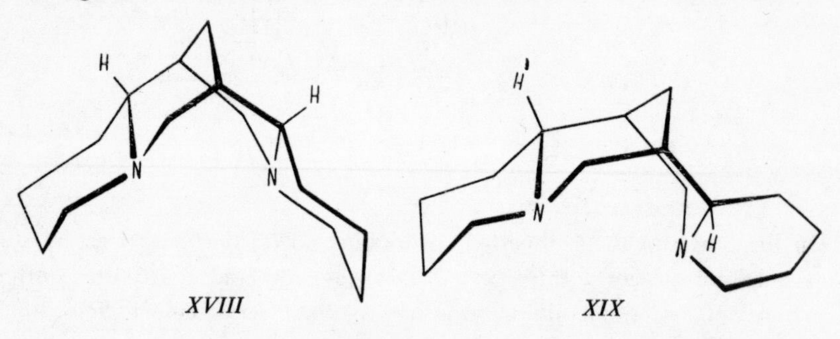

XVIII *XIX*

Besides the factors determining the relative stability of the individual conformers, which we have already referred to, we encounter, in the case of polycyclic compounds, other factors stemming directly from the geometry of such complex molecules. Thus in the case of *trans-anti-trans*-perhydrophenanthrenes (*XX*), substituted at positions 4 and 5, an interaction may occur between the substituents in equatorial position.* In these cases, the lower-energy arrangement will not be the isomer with two equatorial substituents, but the isomer with one equatorial and one axial substituent.

XX

* We may encounter this type of interaction practically in the case of steroid compounds substituted in positions 1 and 11, or in the case of pentacyclic triterpenes.

An important example of the application of conformational analysis to polycyclic systems is provided by the analysis of bond stability in steroid arrangements. It had already been established for both of the basic structures, the 5α-cholestane (*XXI*) and 5β-cholestane (*XXII*) arrangements, that the substituents on the same carbon atom displayed a different reactivity according to whether they were α or β in the sense of the usual Fieser notation. Moreover, the α-isomer is more stable in some cases, the β-isomer in other cases, apparently without regularity. The application of conformational theory brought an explanation. From one aspect this may be regarded as a success for the theory but at the same time we must not forget that the reactions of steroid derivatives with known stereochemistry had enabled formulation of the fundamental rules:

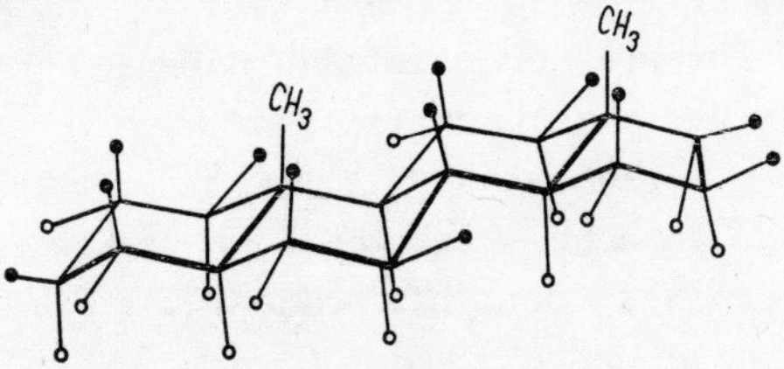

XXI
(5α-Cholestane structure)

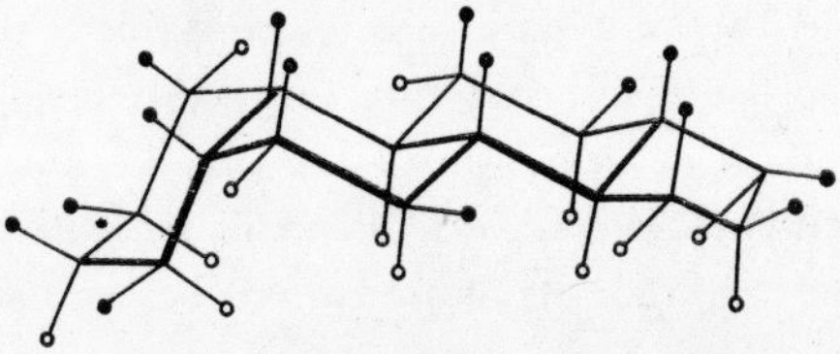

XXII
(5β-Cholestane structure)

In the 5α-cholestane series

2α(e) is more stable than 2β(a)
3β(e) is more stable than 3α(a)
4α(e) is more stable than 4β(a)
6α(e) is more stable than 6β(a)
7β(e) is more stable than 7α(a)

On the other hand, in the 5β-cholestane series

2β(e) is more stable than 2α(a)
3α(e) is more stable than 3β(a)
4β(e) is more stable than 4α(a)

The different stability of the substituents of course also affects their reactivity.

In connection with steroid derivatives, some questions ought to be referred to which are connected with the joining of the five-membered ring D to the six-membered ring C. The five-membered ring by itself is a considerably flatter formation than the cyclohexane ring; its stereochemical structure, as we will see, does not permit us to distinguish bonds of different types which would resemble the equatorial and axial bonds of cyclohexane. The situation is different if the five-membered ring is joined to the perhydrophenanthrene structure of a steroid molecule. Bonds attached to carbon atoms $C_{(15)}$ and $C_{(17)}$ maintain to a considerable extent the character of equatorial and axial bonds. The β-bond attached to $C_{(17)}$ and the α-bond attached to $C_{(15)}$ are then called pseudo-equatorial, and the α-bond attached to $C_{(11)}$ as well as the β-bond attached to $C_{(15)}$ are then called pseudoaxial bonds. Even the bonds attached to the $C_{(16)}$ atom display, to a certain extent, a similiar differentiation, as substituent α is thermodynamically more stable. Conversely, the reverse effect of the five-membered ring on the six-membered one could bring about a certain deformation of the distribution of bonds of the latter, as has been encountered in hydrindane derivatives. However, the rigid perhydrophenanthrene structure cannot be significantly deformed.

Stability of Boat Forms of Cyclohexanes

Four different types of bonds were distinguished in the boat form of cyclohexane, represented by Formula *Va.* The bonds at the carbon atoms $C_{(2)}$, $C_{(3)}$, $C_{(5)}$ and $C_{(6)}$ were designated *be* (*boat-equatorial*) and *ba* (*boat-axial*) because of their resemblance to the normal cyclohexane bonds. The bonds from carbon atoms 1 and 4 are of special character and designated α (*flagpole*) and ε (*bowsprit*). The conformation shown is evidently a very unstable arrangement which could only exist in outstandingly rigid structures. It is comprised of three very high-energy partial conformations: first the double *syn*-periplanar arrangement of bonds in the ring and then the pair of substituents, attached to the bonds designated α in the formula, which come necessarily very close to each other. The maximum of the curve representing the dependence of the potential energy on the rotation around a single bond (see Fig. 10) corresponds to the *syn*-periplanar arrangement. The system will therefore have a tendency to limit the interaction following from this partial conformation by a rotation of a certain angle, the more so because the substituents attached to α-bonds are simultaneously removed from each other (4). The deformation indicated may be performed by rotating the *syn*-periplanar partial conformations in the positive or negative sense.

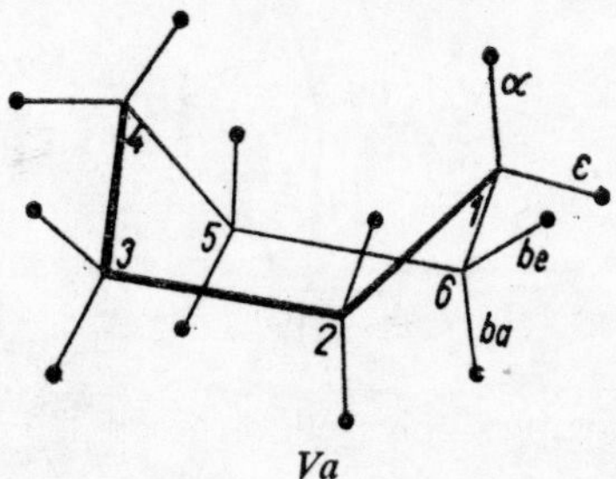

Va

This twisted boat form of cyclohexane has a much higher energy content than the chair form (in total about 5.5 kcal/mol) but, in contrast to the regular boat form, it is a stable structure. The curve representing the

XXIII

change of potential energy in the course of this process incorporates a maximum corresponding to the regular boat, and two minima corresponding to twist boats (skew boats), represented by Formula *XXIII.*

The boat may in some cases become the preferred form or even the only possible one. Its stability may be enforced by the following circumstances.

a) A bridge connecting the 1,4-positions of the cyclohexane ring, either transient (like the hydrogen bond in pseudonortropine or the bridge in the intermediate formed during the migration of the acyl group of the acyl derivatives of the same compound or permanent, as in compounds of the bicyclo[2,2,2]octane and bicyclo[2,2,1]heptane type (*XXIV*).

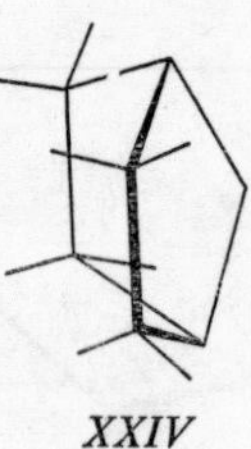

XXIV

b) Inclusion in a complex polycyclic structure, in which the chair form would require the attaching of a second cyclohexane ring by means of two axial bonds (*e.g. trans-syn-trans*-perhydrophenanthrene and *trans-anti-trans*-perhydroanthracene). The boat form thus stabilized was experimentally established for example in the case of 8-isotestosterone (*XXV*).

XXV

c) The presence of substituents which in the chair conformation would, by means of forced localization in the axial position, lead to powerful interaction with other axial substituents. A frequently introduced example of this type is *trans*-1,3-di-t-butyl-cyclohexane which, in a chair conformation, would necessarily have one t-butyl group in an axial position.

It seems however that the existence of such a stabilized **twist-boat form** is not limited to the above extreme and rather curious case. It is possible that a number of other compounds in which we encounter significant 1,3-interactions in the chair form exist partially in the twist-boat form.

Conformations of Partially Unsaturated Cyclohexane Rings

The regularity of distribution of the axial and equatorial bonds on a cyclohexane ring is broken by the presence of an endocyclic double bond fixing atoms $C_{(3)}$, $C_{(2)}$, $C_{(1)}$ and $C_{(6)}$ in one plane. In agreement with the results of spectral measurements we are able to construct either the "half-chair" form (*XXVII*) or the "half-boat" form (*XXVI*). According to calorimetric measurements the former is much more stable. In Formula *XXVII*, there are axial as well as equatorial bonds at the atoms $C_{(4)}$ and $C_{(5)}$. The bonds from the atoms $C_{(3)}$ and $C_{(6)}$ maintain their axial or equatorial character, respectively, only in part; they were termed quasi-equatorial (e') and quasi-axial (a'). The change of position has the effect of increasing the torsional angle contained by the quasi-equatorial bond at $C_{(3)}$ and by the equatorial bond at the vicinal $C_{(4)}$ atom and, conversely, of reducing the torsional angle formed by the corresponding pair of $a'-e$ bonds. In the same manner we may describe the different bonds at the hydroaromatic rings of tetralin and 1,2-epoxy-cyclohexane, where the four carbon atoms are fixed in one plane by attaching another planar ring.

The boat form is the only possible form of the middle cyclohexadiene ring in 9,10-dihydroanthracene.

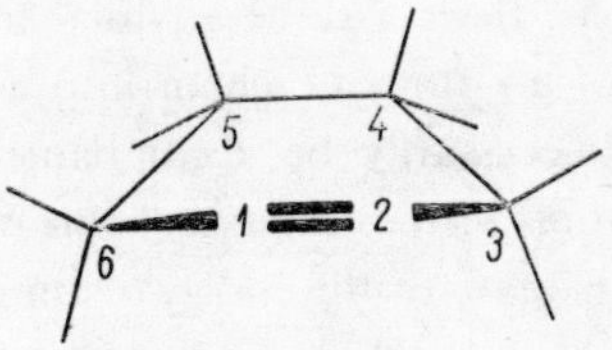

XXVI

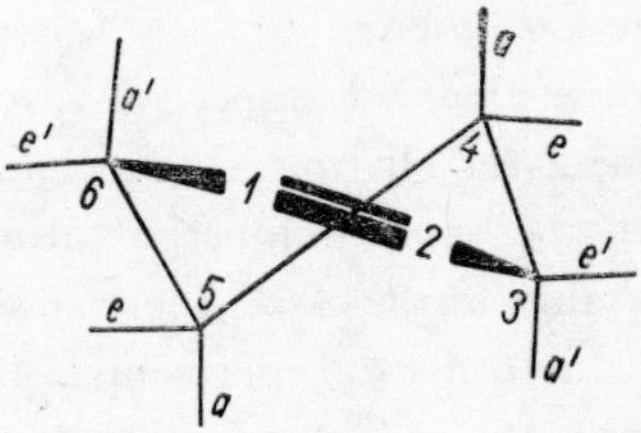

XXVII

Conformations of other than Six-Membered Rings

In the course of conformational analysis we gave the six-membered ring special attention because the required model substances are easily accessible and also because the results are relatively easy to interpret. The several conformations coming into consideration may be derived from the basic conformational states of the tetramethylene chain; for determination of their stabilities the only effective factors are the nonbonded interactions following from rotation around a single bond **(Pitzer strain).** For the conformational analysis of other cyclic structures we have to take further factors into account, which makes the analysis more difficult; for a survey see Reference 3.

Table VIII

DEFORMATIONS OF ANGLES WHICH WOULD BE NECESSARY FOR FORMATION OF PLANAR CYCLOALKANES ACCORDING TO BAEYER'S CONCEPT[a]

Number of methylene groups n	3	4	5	6	7	8	12
$\frac{1}{2}$(Deviation from valence angle 109°28′)	24°44′	9°44′	0°44′	−5°16′	−9°51′	−12°46′	−20°16′

[a] In which the "strain" was equated to $\frac{1}{2}$ (109°28′ — included angle of the regular planar polygon). The factor $\frac{1}{2}$ was used, since the strain is spread over two bonds.

The intrinsic energy content, derived from the heat of combustion measurements, clearly shows the different thermodynamic stability of rings of different size; cyclic structures altogether have a higher energy content than the respective acyclic chains. If we assume a planar arrangement for all rings, then cyclization of an aliphatic chain into a ring with the same number of atoms will necessarily be accompanied by a change of the bond angle at all the atoms of the ring (see Table VIII). This deformation represents a further increase of the energy of the compound (the so-called **Baeyer strain**), which is already large because of the *syn*-periplanar arrangement of all tetramethylene units. In the example of cyclohexane it was demonstrated that by suitable rotation around the bonds we may avoid deformation of the bond angles, at the same time eliminating the nonbonded interactions. The six-membered ring, however, is the only one where such elimination is perfect and where it is not accompanied by other energetically unfavourable interactions, so that the conformation approaches the energy level of aliphatic molecules. The increase of the energy level of the other cyclic structures comprises the following contributions:

a) Pitzer strain, brought about by the nonbonded interactions associated with substituents on adjacent carbon atoms;

b) Baeyer strain, following from the deformation of bond angles;

c) transannular nonbonded interactions, caused by the forced proximity of the opposite parts of the ring.

Small rings have a very high energy content. In the case of a planar three-membered ring, the main energy contribution is provided by the deformation of bond angles, which attains high values in this case. In the instance of a four-membered ring, where the bond angles are also substantially changed, a certain deformation takes place along one of the diagonals, thus modifying the strictly *syn*-periplanar arrangement and lessening the Pitzer strain. In the five-membered ring, the bond angle deformations and, to a certain extent, also nonbonded interactions are eliminated by deflecting one or more of the ring atoms out of the ring plane. Two relatively stable conformations of cyclopentane are thus formed: the first one with four atoms in one plane and the fifth one slightly deflected, resembling an opened envelope (and designated C_s), and the second one in which one atom is deflected below the plane and another one above it, resembling the "half-chair" conformation of cyclohexene (and designated C_2). Both of them may be compared to the conformations of cyclohexene if we imagine that the C=C group is replaced by one methylene group. The Pitzer nonbonded interaction remains the main source of strain in both cases. A seven-membered ring in many cases resembles a five-membered ring, especially in the relative increase of energy compared to cyclohexane and in its behaviour in chemical reactions. The main factor causing the increase of energy, *i.e.* Pitzer strain, is also common to both structures.

Eight- to eleven-membered rings, the so-called **medium rings**, constitute more complex arrangements from the aspect of conformation (5). The negative deformations of bond angles may, to a certain extent, be eliminated by rotation around ring bonds but, on the other hand, considerable nonbonded interactions are introduced. To the usual interactions of the Pitzer type we have to add, in this case, powerful transannular interactions between the two opposite parts of the ring. These interactions are characterized by an increased reactivity during ring opening and more difficult closure of the ring, and they facilitate transannular reactions. Naturally, it is impossible to carry out such a detailed conformational analysis of substances with medium rings as has been executed for cyclohexane derivatives because sufficient data are not yet available. However, the complete shape of the molecule has already been determined for several substances of this series in the crystalline state (with the help of three-dimensional X-ray analysis). The common characteristic of the conformational arrangement of the medium rings hitherto investigated appears to be the overall "S" shape of the molecule rather than the regular "crown" form. The group of five carbon atoms in a double *syn*-clinal arrangement is a recurring element (see *XXVIII*). Formula *XXVIIIa*, representing a cyclodecane conformation established on the basis of an X-ray diffraction study, clearly indicates the presence of this structural unit. The marked hydrogen atoms are atoms displaying considerable nonbonded interaction. Rings with more than twelve

XXVIII *XXVIIIa*

members have a more mobile chain and nonbonded interactions are rather scarce. The properties of the highest members approach the properties of aliphatic substances asymptotically. When performing conformational analysis, we are in these cases more interested in the segment of the ring with the functional groups than in the overall shape of the ring. Such a segment may be analysed practically like the corresponding acyclic chain. The overall conformation represents two parallel chains connected by methylene bridges.

The influence of ring size on the possibility of a cyclization reaction. The three hitherto described general methods of formation of cyclic compounds (*i.e.* the acyloin condensation of dicarboxylic acid esters (*A*), the Ziegler condensation of dinitriles (*B*) and the dry distillation of dicarboxylic acid thorium salts (*C*) display great variation of the yields of cyclization product with variation of the size of the ring. All the methods referred to provide relatively higher yields in the case of five-membered and six-membered rings. In the region of medium rings, the yields fall very rapidly down to zero values and only the first method, acyloin condensation (*A*), is applicable. Twelve-membered and larger rings are again formed much better, especially with the help of methods *A* and *B*. The cause of the significantly small yields of substances with medium rings may be found in the powerful steric interactions taking place in the transition state.

$$(A)\quad \mathrm{ROOC{-}(CH_2)_{\mathit{n}-2}{-}COOR} \xrightarrow{\mathrm{Na}} \overbrace{\mathrm{CO}\text{———}\mathrm{CHOH}}^{(\mathrm{CH_2})_{n-2}}$$

$$(B)\quad \mathrm{NC{-}(CH_2)_{\mathit{n}-2}{-}CN} \xrightarrow{\mathrm{C_6H_5N(Li)(C_2H_5)}} \mathrm{HN{=}}\overbrace{\mathrm{C}\text{———}\mathrm{CH{-}CN}}^{(\mathrm{CH_2})_{n-1}}$$

$$(C)\quad \mathrm{HOOC{-}(CH_2)_{\mathit{n}-1}{-}COOH} \xrightarrow[\text{pyrolytic decomposition}]{\mathrm{ThO_2}} \overbrace{\mathrm{CO}}^{(\mathrm{CH_2})_{n-1}}$$

The influence of ring size on reactivity. The ring strain also affects the size of the rate or equilibrium constants, respectively, of the reactions of cyclic compounds. The comparison of ring size with the stated parameters again divides the cycloalkane homologous series into groups corresponding to the previous classification. The total strain of the ring changes in the course of the reaction. The size of this change, designated as **I strain (Brown strain)**, facilitates the course of the reaction or hinders it, according to the character of the reaction mechanism and especially according to the steric requirements. For an overall assessment of the reactivity of cyclic compounds it is expedient to classify reactions according to the change of the "coordination number"*, which takes

---

* The term represents the change of the number of atoms attached to the respective atom. It is more accurate to speak of the hybridization of valence electrons (sp^3 and sp^2),

Table IX

THE HEATS OF COMBUSTION OF CYCLIC HYDROCARBONS

(Showing the difference per CH_2 group between cyclohexane and the respective cycloalkane)

Ethylene	11.2 kcal/mol	Cyclo-octane	1.2 kcal/mol
Cyclopropane	9.2 kcal/mol	Cyclononane	1.4 kcal/mol
Cyclobutane	6.6 kcal/mol	Cyclodecane	1.2 kcal/mol
Cyclopentane	1.3 kcal/mol	Cyclopentadecane	0.1 kcal/mol
Cyclohexane	0.0		
Cycloheptane	0.9 kcal/mol		

place in the course of the reaction at one of the ring atoms. The conversion of a tetrahedral arrangement with coordination number 4 (sp^3) into a trigonal arrangement with coordination number 3 (sp^2) brings about a change in the bond angles of the ring and especially Pitzer nonbonded interactions. The rings incorporating certain nonbonded interactions in their basic arrangement, for example five-membered and seven-membered rings, display a reduction of nonbonded interactions when a ring atom changes its coordination number from 4 to 3; the reaction will thus proceed very easily. On the other hand, in the case of the six-membered ring with the originally advantageous conformational arrangement the interactions will increase with the introduction of a planar bond system at one of the carbon atoms (for example see 2-alkylcyclohexanones on p. 106). In consequence, the reaction will proceed only unwillingly. The reverse procedure, the change of the coordination number from 3 to 4, encountered for instance in the course of the reduction of cycloalkanones to cycloalkanols, will also display an inverse dependence on the size of the ring. The analogous analysis of medium rings is more difficult with regard to the complexity of the conformation. The reactivity of compounds with large rings approaches that of acyclic compounds. For further details see p. 145.

Effects of Conformation on Chemical Properties

Much more important for organic chemistry than the description of the individual conformers themselves is the difference which will appear in their chemical properties. The different stability of the individual conformational arrangements will assert itself in equilibrium reactions controlled by thermodynamic factors. In the case of reactions influenced by kinetic factors, we have to consider the different steric accessibility of the reaction centre to the reacting molecule, in the transition state.

Epimerization Reactions

An example of the reactions of the first type is given by epimerizations, in the course of which the difference in stability is expressed concretely; the reaction produces a mixture of both epimers in the ratio corresponding to their stability. Thus the action of sodium pentoxide will lead to the epimerization of the axial hydroxyl of tropine (*XII*) into the equatorial hydroxyl of pseudotropine; during the mutarotation of sugars, an equilibrium controlled by conformational factors is established between the α and β pyranose and furanose forms. The effect encountered with ketones of the perhydrophenanthrene series is interesting, where a change of the way the rings are connected takes place in the course of epimerization (see Diagram C 3). Whilst isomerizations *A* and *B* proceed smoothly, reaction *C* cannot be performed at all under the same conditions. The isomer which would thus be formed would necessarily have the middle ring in a boat form. The epimerization is facilitated by the presence of an adjacent carbonyl group. Even a hydrocarbon which is not activated may be subjected to epimerization under suitable reaction conditions. Thus 1,4-dimethylcyclohexane yields a mixture containing more than 90 % of the *trans*-isomer when treated with 98 % sulphuric acid.

Reactivity of Groups in Axial and Equatorial Positions

The determination of the difference in reactivity between an axial and equatorial group requires the possibility of executing a reaction with isomers having a guaranteed stable conformational position of the substituent. Best suited to this purpose are cyclohexane derivatives substituted with a t-butyl group at the non-vicinal carbon, as well as, to a lesser extent, steroid derivatives in which their rigid polycyclic structure prevents isomerization between conformers. In these derivatives the reactivity of a certain substituent may be affected by the action of other substituents which may be located even in a distant part of the basic arrangement.

In general it may be said that equatorial substituents are more easily attacked by reagents and react more readily in the case of reactions

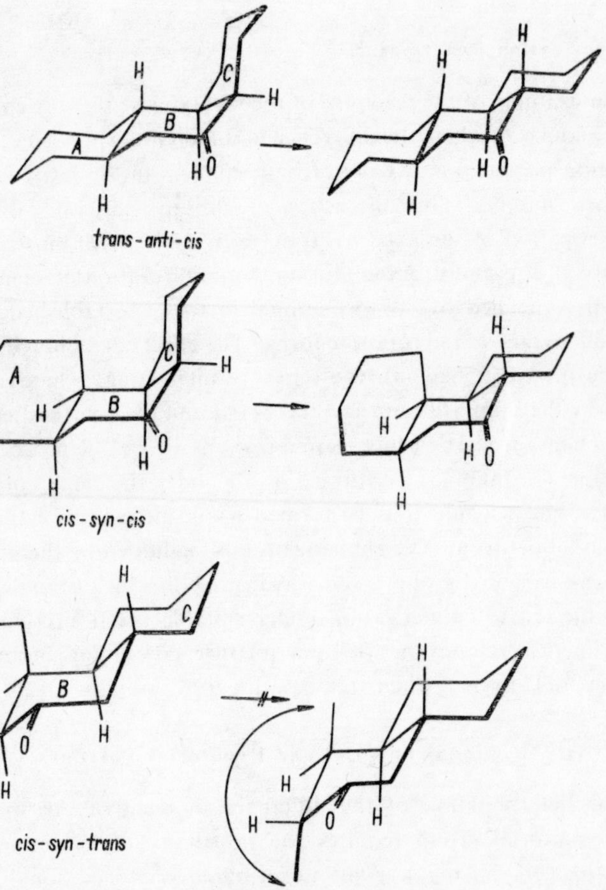

trans-anti-cis

cis-syn-cis

cis-syn-trans

Diagram C 3*

---

* The configurational prefixes indicate subsequently the relative position of hydrogen atoms at the atoms $C_{(5)}-C_{(10)}$ (junction of rings *A* and *B*), $C_{(10)}-C_{(9)}$ and $C_{(9)}-C_{(8)}$ (junction of rings *B* and *C*). The last reaction does not occur because ring *A* would have to be fused to ring *B* by means of two axial bonds, which is not practicable.

occurring directly on them. In agreement with this statement, the esterification of equatorial hydroxyl and carboxyl groups proceeds at a much greater rate than that of the analogous axial groups (see Table X). Very expressive stereospecificity is also displayed by the quaternization of the tertiary amino group. Compounds with an equatorial dimethylamino group (*e.g.* 3β-dimethylamino-5α-cholestane or 3α-dimethylamino-5β-cholestane) are characterized by a lower activation energy of the reaction with methyl iodide, and their reaction rates are much higher than those of their $C_{(3)}$ epimers. The fact that the acids as well as bases with acidic or basic axial groups, respectively, are all weaker than in the case of the equatorial isomers may also be best explained by the more difficult solvation of the less accessible axial groups.

Table X

RELATIVE RATE OF ESTERIFICATION OF HEXAHYDROPHTHALIC ACIDS

Isomer	*Conformation*	*Relative rate of esterification*	
		1st carboxyl	*2nd carboxyl*
1,4-*cis*	*e, a*	1.6	0.067
-*trans*	*e, e*	2.8	1.32
1,3-*cis*	*e, e*	3.0	1.18
-*trans*	*e, a*	1.7	0.07
1,2-*cis*	*e, a*	0.17	0.03
-*trans*[a]	*a, a*	0.078	0.045

[a] The results show that neighbouring, strongly polar groups are stabilized predominantly in diaxial conformations (in the *trans*-isomer).

On the other hand, in the case of reactions in which a substituent is split off, the axial isomer reacts more readily because the elimination of the substituent means also the elimination of a 1,3-interaction between it and further axial substituents on the same side of the cyclohexane ring.

By the removal of the axial substituent and by the elimination of the 1,3-interaction we may also explain the more rapid oxidation of steroid axial hydroxyl groups by chromium trioxide.

Table XI

OXIDATION OF ISOMERIC 5α-CHOLESTANOLS

Substituent position		1	2	3	4	6	7
Relative rate of	*e*	9.7	1.3	1.0	2.0	2.0	3.3
hydroxyl oxidation	*a*	13.0	20.0	3.0	35.0	36.0	12.3

Table XI shows that all equatorial hydroxyls are oxidized more slowly and that the rate of oxidation is almost the same for all equatorial hydroxyls. The exclusive position of the 1*e*-hydroxyl group is due to its interaction with the equatorial hydrogen at $C_{(11)}$. With the axial hydroxyl groups, the rate varies somewhat, more or less in agreement with the intensity of their 1,3-interactions. The elimination of 1,3-interactions will also have an effect upon the rate of substitution reactions proceeding by S_N2 mechanisms. The axial toluene-*p*-sulphonyloxy group in *cis*-4-t-butylcyclohexyl toluene-*p*-sulphonate is replaced by the thiophenoxide anion 19 times more rapidly than the equatorial group of the *trans*-isomer. Walden inversion takes place in both cases, as is usual with S_N2 reactions. However, the substitution of the axial group is accompanied by considerable elimination, so that the *cis*-isomer yields 42% of *trans*-4-t-butylcyclohexyl phenyl sulphide and also 32% of 4-t-butylcyclohexene.

Formation of New Asymmetric Atoms

The access of a reagent to a reaction centre is influenced by the substitution in the vicinity of the latter. If the molecule has a chiral structure, as is usual for cyclohexane derivatives, the access of the reagent to the molecule is easier from one side than from the other and the two transition states thus display, on the basis of nonbonded interactions, different strain, so that the final products are formed from them at different rates and in different quantities. The stereospecificity of these reactions is considerable in the cyclohexane series. An illustrative example is the reduction of substituted cyclohexanones to the respective hydroxy derivatives. Catalytic hydrogenation leads, as a rule, to the derivative with an axial hydroxyl, especially when executed in an acidic medium. On the other hand, the use of metallic reducing agents or the direct use of metals (and also the addition reactions of organometallic compounds) enable us to prepare compounds with an equatorial hydroxyl, in agreement with the reaction mechanism of both types of reduction (for details see p. 203).

Effect of Conformation on Formation of Cyclic Compounds

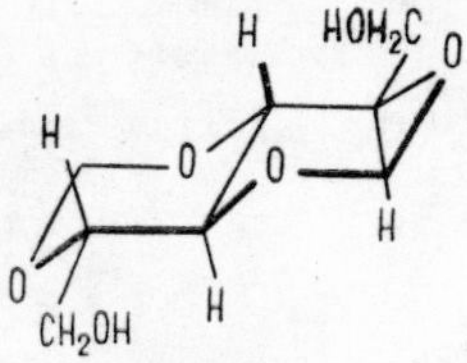

XXIX 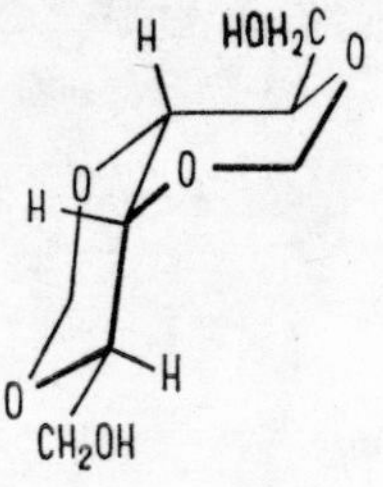 *XXX* 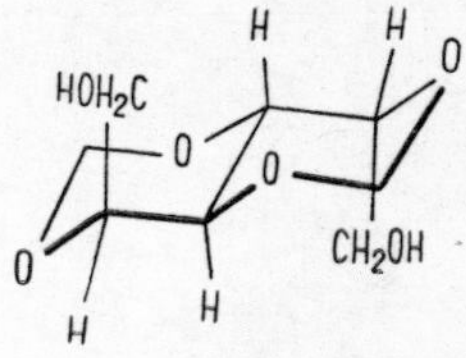 *XXXI*

The principles of conformational analysis may be applied to heterocyclic compounds. For example the relative stabilities of fused rings in diacetals of hexitols may be predicted. An example of the most stable type of acetal is 2,4: 3,5-di-*O*-methylene allitol, which has a *trans*-fusion of the rings and both substituents in equatorial positions. From the reaction of 1,6-di-*O*-benzyol-D-mannitol with formaldehyde followed by debenzoylation was obtained 2,4: 3,5-di-*O*-methylene-D-mannitol. It can fulfil the requirement of equatorial orientation of the substituents only in the higher energy *cis*-fused form (*XXX*). The isomeric derivative of galactitol (*XXXI*) could not be prepared because it would have to have both substituents axial in the *trans*-fused form.

Destabilizing conformational factors exert a marked influence also upon reverse reactions, when ring opening takes place. Of the isomeric tetrahydrooxazines (*XXXIII*) and (*XXXIV*), obtained from the *erythro*- and *threo*-1,3-diphenyl-3-aminopropanols (*XXXIIa*) and (*XXXIIb*), the second one (with one phenyl group in an axial position) is cleaved a hundred times more quickly than the first one with both phenyls in equatorial positions.

Finally, we have to note the ease with which the two bonds from neighbouring cyclohexane atoms may be made coplanar. The two bonds from neighbouring cyclohexane atoms (either two equatorial or one axial and one equatorial) contain a torsional angle of 60° in an ideal chair conformation. If the two substituents attached to them are to become components of a five-membered ring, then this angle has to be reduced, thus necessarily destroying the regularity of the chair form.

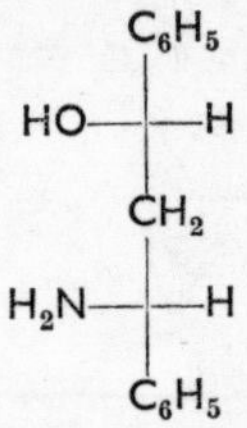

erythro-
XXXIIa

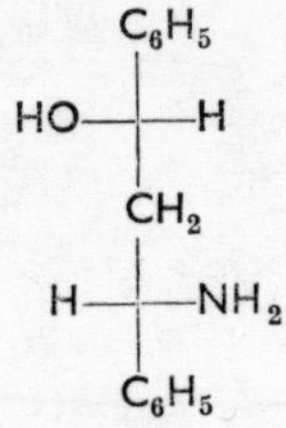

threo-
XXXIIb

The attempt to make the two equatorial bonds coplanar brings about a significant deformation of the ring (increased puckering) and an increase of internal strain. In the case of the axial equatorial pair of bonds the whole ring is flattened simultaneously with the reduction of the angle contained by the bonds and the increase in internal strain is smaller (Formulas *XXXV* and *XXXVI*).

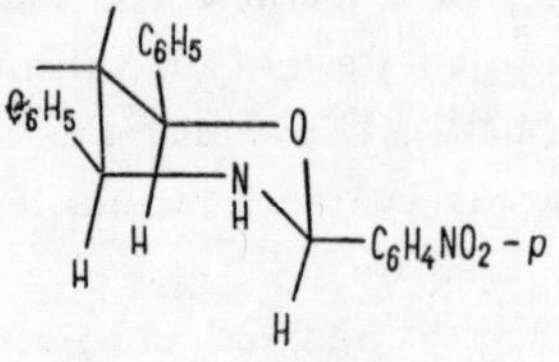

XXXIII

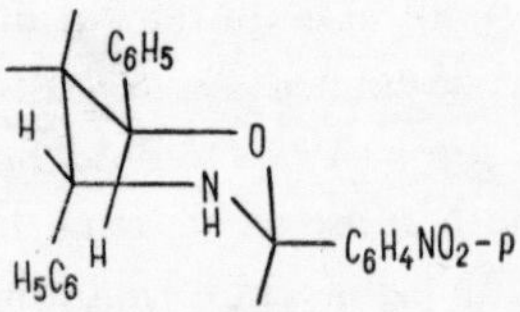

XXXIV

XXXV

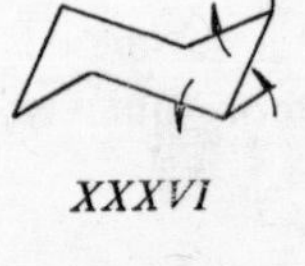

XXXVI

Effects of Conformation on Physical Properties

The different steric structure appearing in conformers is bound to influence their physico-chemical properties. As has already been said, it is necessary to revise the original wording of the Auwers-Skita rule in the case of 1,3-disubstituted cyclohexanes. The rule enables the determination of the probable relative configuration on the basis of simple physical properties of substances. The difference between the equatorial and axial position of a substituent may, in some cases, exert a very marked influence upon its spectra. For instance, the characteristic infra-red absorption band of the carbonyl group in α-bromoketones is shifted by 20 cm^{-1} in the direction of higher frequencies compared with the corresponding band of the non-brominated ketone, if the bromine is in an equatorial position. On the other hand, the bromine atom in an axial position practically does not cause any shift at all. Characteristic differences were also encountered in the case of steroid alcohols and their acetates.

In the course of paper or column (aluminium oxide) chromatography, compounds with axial hydroxyl groups travel more rapidly than the isomers with an equatorial hydroxyl group. The equatorial hydroxyl group evidently facilitates adsorption on the surface of the adsorbent.

Principal Physico-Chemical Methods Used for Conformational Studies

a) *Electron diffraction measurements* were of principal importance for the first investigations by Hassel. They enable the direct localization of atoms and the determination of interatomic distances. As measurements are carried out in the gaseous phase, intermolecular attractive or repulsive forces will have no effect. These forces may influence the results of other methods very unfavourably.

b) *X-ray diffraction measurements* with solid substances enable us to determine the conformation of the substance in the solid state. The conformation is stabilized by the action of all the surrounding molecules in the crystalline lattice and may differ from the conformation which the molecule would occupy by itself. The conclusions drawn from these measurements therefore have to be applied carefully to substances in solution.

c) *Nuclear magnetic resonance*, as the only method directly detecting the position of the nuclei of hydrogen atoms and their common position, proved to be outstandingly suited to the study of stereochemical problems. In the field of conformational analysis it enables the determination of the partial conformation at certain bonds (*e.g.* the presence of an axial or equatorial hydrogen atom or the way two neighbouring rings are connected) even if the molecular segment is of the hydrocarbon type.

d) *Optical rotatory dispersion*, especially in the region of the Cotton effect, permits the determination of the partial conformation in the vicinity of an optically active chromophore. Comparison with model substances is not necessary when using the octant rule. However, the method is applicable only for optically active compounds with a characteristic absorption in the accessible ultra-violet region of the spectrum. Its special importance is in the investigation of the secondary structure of macromolecular substances. The application of *circular dichroism measurements* is analogous.

e) *Dipole moment measurements* enable us to select certain conformations from a set of others under the assumption of being able to calculate the respective dipole moment of each. For purposes of conformational analysis the only applicable moments are those of exactly orientated groups, *i.e.* those which act in the direction of the bond (*e.g.* C—Cl or C=O, but not C—O—H). By limiting the number of possible conformations we get a valuable guide for further investigations.

f) *Thermodynamic calculations*, comparing the entropy calculated from other thermodynamic data with the entropy calculated from spectroscopic data, are applicable only for the simplest compounds. They played an important role in the deriving of the basic energy parameters characterizing the individual states of rotational isomerism.

Conformation of Macromolecules

If we would attempt to broaden the method of conformational analysis as described above to include macromolecules, we would be up against a difficult task of choosing the most stable conformation from an immense number of other possible conformations. Fortunately, natural as well as synthetic macromolecules accommodate periodically repeated structural units and such structural regularity is then reflected by the regularity of spatial arrangement as well. The interactions between structural units are the same or at least very similar throughout the

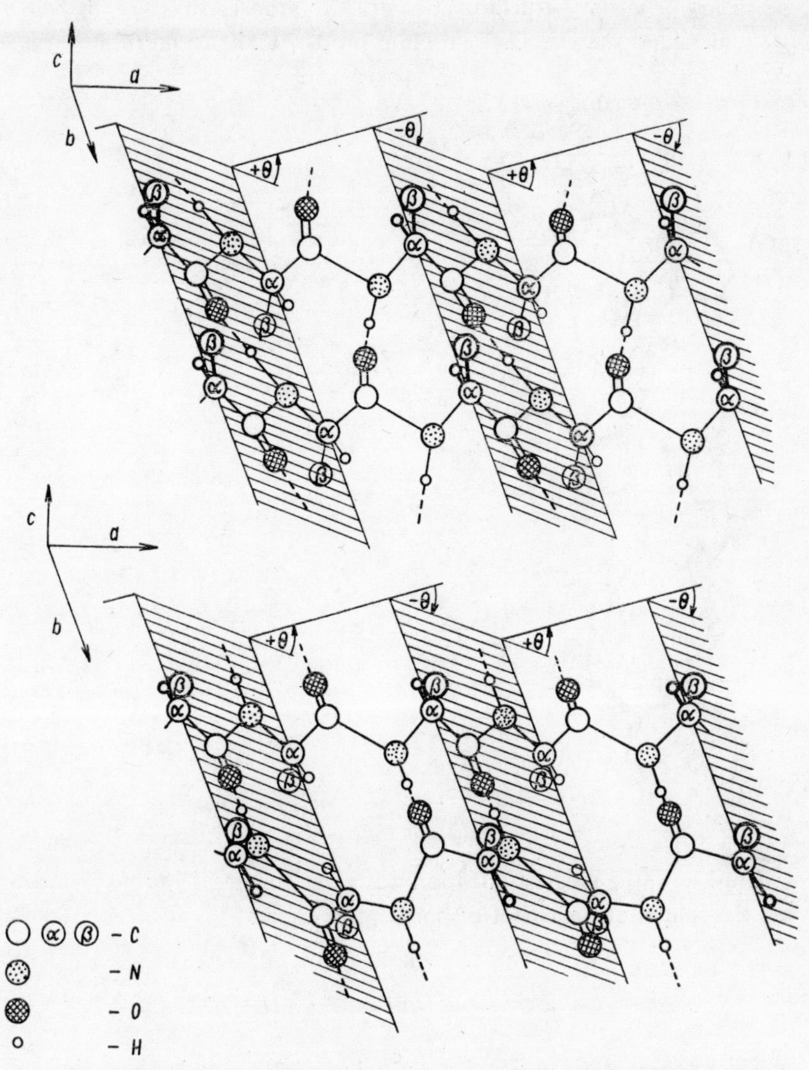

Fig. 12. Diagram for the pleated sheet structure (β-structure).

chain and, under suitable conditions, provide the chain with a particular shape. In the range of natural macromolecular substances, we often do not speak of conformation in this connection but we use the term **secondary** or **tertiary structure**, in contrast to **primary structure**, by which we understand the chemical structure proper of the chain (6).

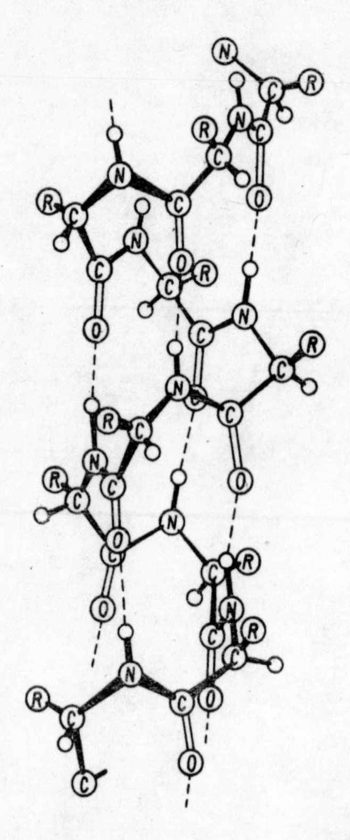

Fig. 13. α*-Helix.*

Until now, the greatest attention was given to considerations concerning the spatial arrangement of polypeptide chains

$$-NH-\overset{\overset{R_1}{|}}{C}H-CO-\left[-NH-\overset{\overset{R_2}{|}}{C}H-CO-\right]_n-NH-\overset{\overset{R_3}{|}}{C}H-CO-,$$

in which the regularly repeated structural unit is the α-amino acid segment (the part of the formula within brackets). With regard to the possible mesomerism of the peptide bond, the bond between the carbonyl carbon atom and the nitrogen atom has, in part, the character of a double bond and the six atoms (bold print) are contained in one plane. The planar segments may then be arranged in a different manner (7). The best-known are Pauling's so-called β-**structures** (pleated sheet), in which the peptide chains, with all bonds approximately in *anti*-periplanar conformation, are situated in parallel layers arranged above one another, forming a three-dimensional system (Fig. 12), and the so-called α-**helix**, where the peptide chain forms a helix as shown in Fig. 13. The interatomic forces holding these conformations cannot, of course, be designated as nonbonded repulsive interactions only. Without doubt, an important role is played by lipophilic (hydrophobic) interactions, holding close the hydrocarbon side chains of the amino acids, as well as hydrogen bonds, for example the interpeptidic ones, due to the polar character of the peptide.

The helical array seems to be a very general type of secondary structure, not only in the case of natural macromolecules (biopolymers) such as proteins or nucleic acids, but also in the instance of synthetic polymers (*e.g.* olefin polymers) and even in the case of inorganic polymers (*e.g.* the fibrous modification of sulphur). The helices coming into consideration for the individual types of substances of course differ considerably as regards their geometrical parameters (coil distance, diameter, *etc.*) as well the character of interatomic forces stabilizing them.

Literature

1. Barton D. H. R., Cookson R. C.: Quart. Rev. *10*, 44 (1956).
2. Hanack M.: *Conformation Theory*. Academic Press, New York, 1965.
3. Eliel E. L., Allinger N. L., Angyal S. J., Morrison G. A.: *Conformational Analysis*, Wiley, New York, 1965.
4. Balasubramanian M.: Chem. Revs. *62*, 591 (1962).
5. Sicher J.: *Progress in Stereochemistry*, (P. B. D. de la Mare, W. Klyne, Eds.) Vol. 3, p. 202. Butterworths, London, 1962.
6. Bláha K.: Chem. Listy *58*, 1064 (1964).
7. Pauling L., Corey R. B.: *Fortschritte der Chemie organischer Naturstoffe*, Vol. XI. p. 180. Springer, Vienna, 1954.

II. DYNAMIC STEREOCHEMISTRY

D. STEREOCHEMICAL COURSE OF SUBSTITUTION REACTIONS

General Comments on Mechanism

In the course of substitution reactions, atoms or groups of atoms are replaced by other atoms or groups. The mechanism of these reactions may vary; if a dissociation into ions or radicals precedes the formation of a new bond it will differ from that when the reaction takes place in such a manner that a bond is broken on one side and simultaneously a new bond is formed on the other side.

Before starting to consider the individual types of substitution reactions, we have to elucidate, when discussing the reaction mechanism, which of the two reacting molecules is the **reagent** and which is the substrate with which the reaction is executed. It has become usual to consider as the reagent the simpler substance giving similar reactions with a number of organic compounds. Under the assumption that there are no doubts as to which of the reacting substances is the reagent and which is the substrate, we classify reactions into groups according to the **character of the reagent**. The reagents which in the course of the reaction transfer electrons to another atom are called **nucleophilic reagents**. Conversely, the substances which in the course of bond formation gain electrons from another molecule are called **electrophilic reagents**. Consequently, nucleophilic reagents have at least one free electron pair available, whether they carry a charge or whether they are electrically neutral.

Examples of nucleophilic reagents:

Br^-, Cl^-, I^-, HO^-, RO^-, RS^-,
CN^-, $RCOO^-$, $(ROOC)_2CH^-$, H_2O,
ROH, NH_3, RNH_2, R_2NH, R_3N, N_3^-,
H^-, $RMgX$, RLi *etc.*

Examples of electrophilic reagents:

$$Cl^+, Br^+, I,^+ NO_2^+, AlCl_3, BF_3, \textit{ etc.}$$

As regards the heterolytic cleavage of bonds, we distinguish nucleophilic substitution (S_N) and electrophilic substitution (S_E), according to the character of the the attacking substance:

$$Y: + R \mid - X \rightarrow Y - R + X: \quad (S_N)$$

$$Y + R - \mid X \rightarrow Y - R + X \quad (S_E)$$

In the first case, the bond in the newly formed compound was formed of electrons which belonged to the nucleophilic reagent. The reaction should correctly be called substitution by means of a nucleophilic reagent, however, for the sake of brevity we designate it as **nucleophilic substitution.** In the second case, the electron pair for the newly formed bond was brought in by the molecule on which we carried out the substitution. In this case we speak of **electrophilic substitution.** Nucleophilic substitution reactions are characteristic for aliphatic and alicyclic compounds whereas electrophilic substitution prevails in the case of aromatic compounds. To the stated types of substitution reactions we have to add reactions in the course of which bonds are cleaved homolytically (radical mechanism):

$$Y\cdot + R\cdot \mid \cdot X \rightarrow Y - R + X\cdot \quad (S_H)$$

The classification of substitution reactions into nucleophilic and electrophilic ones gives us a picture of the initial and final states of the reaction only. The total change of the system is, however, not determined by these limiting states but is very closely connected with the mechanism of the reaction. Therefore, we classify reactions further according to the number of particles (ions or radicals), the interaction of which leads to the chemical conversion, which expresses the molecularity of the reaction. The molecularity of the reaction always refers to the reaction step determining the overall reaction rate and, in this step, corresponds to the number of molecules where the change of a covalent bond takes place.

Substitution reactions may occur by means of two mechanisms. With a unimolecular mechanism the compound first of all dissociates into ions or radicals and the rate of their formation at the same time determines the rate of the whole reaction. In the case of a unimolecular nucleophilic substitution (S_N1), which will interest us primarily, a cation is formed by dissociation which then quickly reacts with an anion or a neutral molecule forming the product. If it reacts directly with a molecule of the solvent we as a rule speak of a solvolytic substitution reaction. This includes the reactions of alkyl halides and esters of toluene-*p*-sulphonic acid or of *p*-bromobenzenesulphonic acid with the solvent, which may be represented by water, alcohols, aqueous alcoholic solutions organic acids, *etc.* However, more detailed investigation of unimolecular reactions established that the formation of a carbonium ion does not occur so simply, but that the dissociation of the molecule proper into independent ions is preceded by ionization to a pair of ions which are associated by electrostatic attraction, which is then disrupted by dissociation proper with the participation of solvent molecules solvating both ions. This generalization indicates the importance of the solvent, especially if it is strongly polar. Unimolecular S_N1 reactions are always accompanied by elimination or possibly by rearrangement of the carbon chain (see p. 169).

From the stereochemical point of view it is important to note that carbonium ions, which are virtually planar structures, are intermediates in S_N1 reactions. The result of a substitution reaction on an optically active compound with an asymmetric carbon atom directly taking part in the reaction is, in an idealized case, the formation of a racemate, because the nucleophilic particle may gain access to this ion from both sides with the same probability (see Diagram D 1). Besides predominant racemization, a partial inversion of the configuration may take place. The polar solvent enhances the ionization of the C—X bond. If the solvent is capable of coordination with the displaced particle X, increased solvation will take place. The result of the interaction is the formation of a so-called **solvated ion pair** resembling the transition state of bimolecular substitution reactions. If the nucleophile Y enters before the free, symmetrically solvated ion is formed, we obtain a product of o pposite configuration:

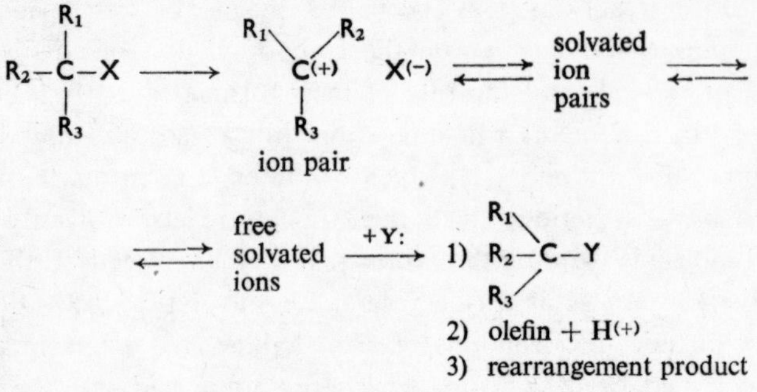

$$R_2{-}\overset{R_1}{\underset{R_3}{C}}{-}X \longrightarrow \underset{\text{ion pair}}{R_1R_2R_3C^{(+)}\quad X^{(-)}} \rightleftarrows \text{solvated ion pairs} \rightleftarrows \text{free solvated ions} \xrightarrow{+Y:}$$

1) $R_1R_2R_3C{-}Y$

2) olefin + $H^{(+)}$

3) rearrangement product

Diagram D 1

In the opposite case, the product of the substitution is a racemate (see Diagram D 2). A special case may ensue if the substrate contains neighbouring atoms capable of playing the role of a nucleophilic particle.

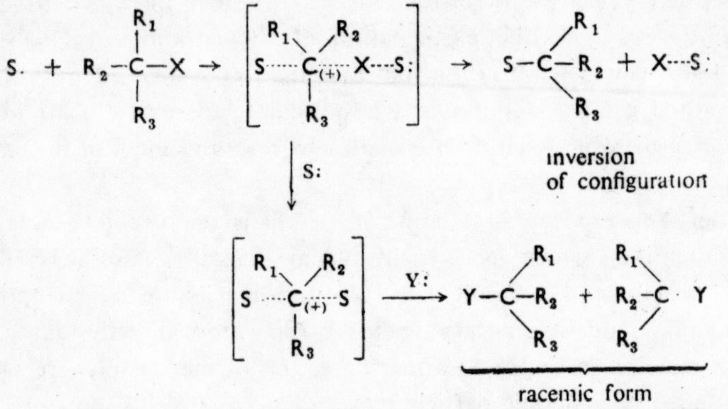

$$S. + R_2{-}\overset{R_1}{\underset{R_3}{C}}{-}X \rightarrow \left[S\cdots R_1R_2R_3C_{(+)}\cdots X\cdots S{:}\right] \rightarrow S{-}C(R_1)(R_2)(R_3) + X\cdots S{:}$$

inversion of configuration

$$\left[S\cdots R_1R_2R_3C_{(+)}\cdots X\cdots S{:}\right] \xrightarrow{S:} \left[S\cdots R_1R_2R_3C_{(+)}\cdots S\right] \xrightarrow{Y:} \underbrace{Y{-}C(R_1)(R_2)(R_3) + R_1R_2R_3C{-}Y}_{\text{racemic form}}$$

Diagram D 2

For example, if such an atom is in a vicinal position to the carbon atom at which the substitution takes place, and if the substrate is capable of occupying an *anti*-periplanar arrangement, the neighbouring atom will assist ionization and increase the reaction rate of the whole reaction and also ensure that the carbonium ion cannot occupy a planar configuration, so that the substitution will proceed with retention of configuration (Diagram D 3). The stated course of the reaction is similar to the mechanism of a number of nucleophilic rearrangements, where particle Z migrates together with the pair of bonding electrons to the neighbouring carbon atom. Besides these unimolecular substitution reactions we have intramolecular substitution reactions (S_Ni) in which a cyclic mechanism takes part.

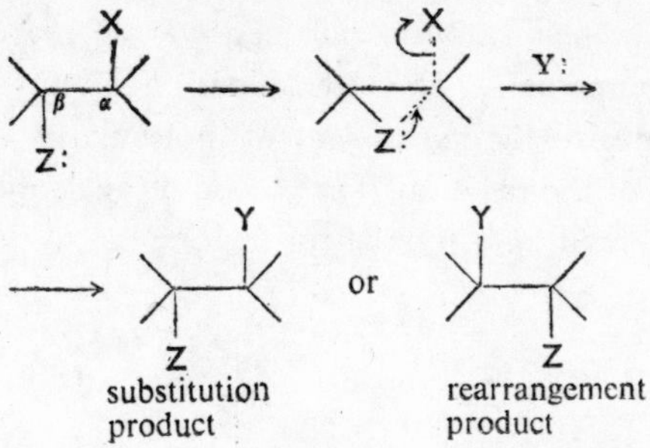

substitution product rearrangement product

Diagram D 3

The stereochemical course of unimolecular electrophilic (S_E1) or radical (S_H1) reactions is identical, even though in the first case we encounter carbanions during substitution whereas the second case involves radicals with one unpaired electron. Without discussing in greater detail the whole course of these reactions we may say that all of them are accompanied by a racemization of the product, caused by the instability of the configuration of the carbanions and of the free radicals. The configurational stability of saturated carbanions may be compared to the configurational stability of tertiary amines (see p. 27). If we find in the literature cases of the retention of configuration in the course of S_E1 reactions, these reactions are mostly carried out at extremely low temperatures (−80 °C), when the carbanion can maintain its original configuration with regard to the limited motion of the molecules. To an even greater extent, the same is true of S_H1 reactions.

In the case of bimolecular substitution reactions, a new bond is formed simultaneously with the rupture of the original one. S_N2 reactions are started by a collision, during which nucleophilic particles approach the substrate from the opposite side to the leaving group, because in this case the repulsion of the two particles is the least. The three remaining bonds are thus inverted, in a similar manner to an umbrella in the wind. In the instant of substitution, in the transition state, the bonds which do not take part in the reaction are contained in one plane. From the stereochemical aspect all S_N2 reactions are accompanied by a change of configuration at the reacting centre, *i.e.* by the so-called **Walden inversion** (Diagram D 4). The S_N2 reactions are, as a rule, reactions of the second order and their reaction rate depends on the concentration of both reacting components. The determination of this dependence frequently contributes to the assessment of molecularity and thus also of the steric source of the reaction. The course of nucleophilic substitution reactions is influenced by a number of factors.

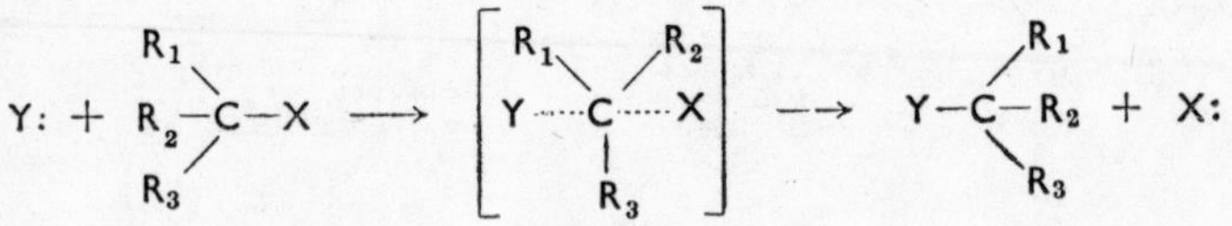

$$Y: + R_2{-}C(R_1)(R_3){-}X \longrightarrow \left[Y \cdots C(R_1)(R_2)(R_3) \cdots X \right] \longrightarrow Y{-}C(R_1)(R_3){-}R_2 + X:$$

transition state

Diagram D 4

1. It depends first of all on the structure of the attacked molecule. The tendency of molecules to react by means of a unimolecular mechanism is connected with the ease of formation and stability of the carbonium cation, which is larger the greater the branching of the carbon chain at the carbon atom carrying the positive charge. The bulky alkyl groups attempt to pass from the tetrahedral arrangement to the lower-energy planar arrangement thus helping, in addition to electronic factors, the dissociation of the substituted particle. (The formation of allyl- and benzyl-type cations is easy, because the carbon sextet is conjugated with the system of π-bonds.) On the basis of these assumptions we may list the approximate sequence of alkyl groups expressing their capability to react by S_N1 or S_N2 mechanisms:

$$(C_6H_5)_3C{-} > (C_6H_5)_2CH{-} > C_6H_5CH_2{-} =$$

$$= R{-}CH{=}CH{-}CH_2{-} > R_3C{-}CH_2{-} > R_3C{-} > R_2CH{-} > RCH_2{-}$$

$$\longleftarrow S_N1 \longrightarrow \quad S_N2 \longrightarrow$$

The sequence is to be understood in such a manner that the substitution typical for the alkyls to the left is unimolecular substitution, whereas bimolecular substitution is less frequent. In the case of the alkyls to the right the situation is the reverse. Both reaction mechanisms are thus possible, especially in the case of the alkyls located in the centre of the above series. For example, compounds of the neopentyl type $(CH_3)_3C—CH_2—$ are unusually unreactive in the course of S_N2 reactions. However, if they react by an S_N1 mechanism, their reaction rate is analogous to other types of S_N1 reactions. Especially inert in S_N1 as well as S_N2 reactions are the derivatives of bicyclic compounds with the reacting substituent located at the junction of the rings.

2. The polarity of the solvent is very important for all types of substitution reactions. The greater the polarity of the solvent, the more it supports the S_N1 reaction. In the case of S_N2 reactions polar solvents support the reaction between electrically neutral molecules (for example the reaction of amines with alkyl halides) and, on the other hand, slow down the reaction between ions and neutral molecules (the saponification of alkyl halides, the halogen exchange reactions in alkyl halides) or even between two ions (the cleavage of quaternary ammonium bases).

3. The course of the substitution is influenced by the nucleophilic character of the attacking particle as well as by the character of the displaced group.

We will demonstrate the validity of these statements using several examples. In the course of solvolytic reactions occurring by an S_N1 mechanism, racemization takes place most readily in the case of biphenylmethanol derivatives and similar compounds. Similarly, optically active α-phenylethyl chloride is also easily racemized; in the course of its solvolytic reaction in methanol and ethanol the ether formed is usually racemic; in 60 % and 80 % acetone a 50 % and 20 % inversion of the configuration has been observed. The change of configuration is encountered to a great extent in the case of S_N1 reactions of secondary alkyl halides. A 70 % inversion was observed in the course of the solvolytic reaction of 2-bromooctane in 60 % ethanol. Primary alkyl halides react almost exclusively by an S_N2 mechanism; the Walden inversion was proven in this case with the help of deuterated compounds. Inversion takes place almost always in the course of the reaction of alkyl halides with hydroxyl, acetate, cyanide and azide ions.

The esters of sulphuric acid and the esters of sulphonic acids react in a similar manner. Conversely, we obtain esters from the reaction of acyl and alkylsulphonyl chlorides with alcohols, in which the original configuration of the alcohol is maintained, because the reaction occurs altogether at the oxygen atom and the asymmetric carbon remains untouched (Diagram D 5). The reduction of optically active alkyl halides by lithium aluminium hydride occurs probably by an S_N2 mechanism, *i.e.* with a change of configuration. Optically active α-deuterioethylbenzene has been prepared by the reduction of α-chloroethylbenzene with $LiAlD_4$.

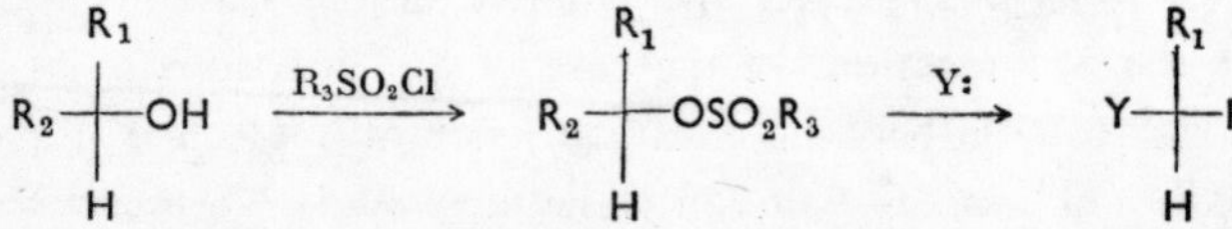

$$R_2\text{—}C(R_1)(H)\text{—}OH \xrightarrow{R_3SO_2Cl} R_2\text{—}C(R_1)(H)\text{—}OSO_2R_3 \xrightarrow{Y:} Y\text{—}C(R_1)(H)\text{—}R_2$$

$Y: = RCOO^{(-)}, HO^{(-)}, Cl^{(-)}, CN^{(-)}, N_3^{(-)}$ etc.

Diagram D 5

The reactions of optically active alkyl halides or sulphonyl esters of optically active alcohols with tertiary amines produce quaternary ammonium salts with opposite configuration. The reaction with dialkyl sulphides is also S_N2. The manner in which the reaction rate is affected by the presence of various groups in the vicinity of the nitrogen atom may be observed from the ease of quaternization of *N,N*-dimethylaniline, *N,N*-dimethyl-2,6-xylidene and *N,N*-dimethyl-2-isopropylaniline, which decreases in the order shown. Brown *et al.* studied the influence of the bulk of the alkyl group in the α-position of a pyridine ring and of the bulk of the alkyl halide reagent on the formation of quaternary salts. The comparison of reactivity of alkyl halides with triethylamine and quinuclidine is interesting. Although both these bases are of the same order of strength, the accessibility of the nitrogen atom is increased by preventing the rotation of the alkyl groups attached to the nitrogen as the result of the fusion of another ring in the case of quinuclidine, so that a lower activation energy is sufficient for the formation of quaternary salts.

The decomposition of quaternary ammonium salts and bases is also accompanied by Walden inversion, if elimination does not take place. For example, the pyrolysis of optically active (α-phenylethyl)trimethylammonium acetate (*I*) yields the acetate of α-phenylethyl alcohol (*II*) with 98 % inversion. In the case of (−)-piperityltrimethylammonium hydroxide (*III*), where the conditions of hydrogen β-elimination are not fulfilled, cleavage produces (+)-neopiperitol (*IV*) only, again with inversion of configuration. Esterification and saponification may formally be considered as "the substitution of a hydrogen by an alkyl group and *vice versa*". The difficult esterification of a number of aliphatic acids

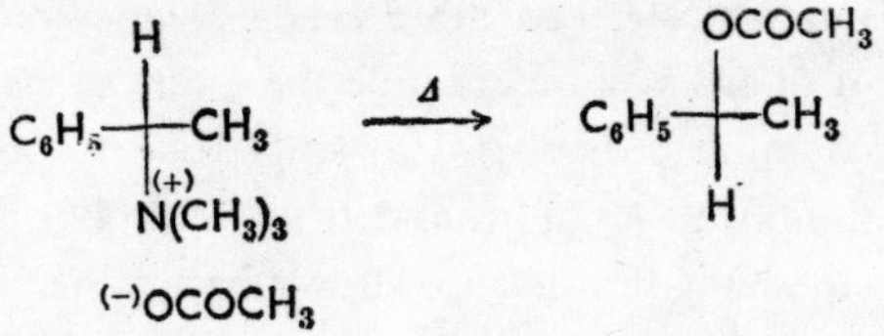

H
C_6H_5—C—CH_3
(+)
$N(CH_3)_3$
(−)$OCOCH_3$
I

$\xrightarrow{\Delta}$

$OCOCH_3$
C_6H_5—C—CH_3
H
II

CH_3
H
(+)
$N(CH_3)_3$
H
(−)OH
III

$\xrightarrow{\Delta}$

CH_3
OH
H
H
IV

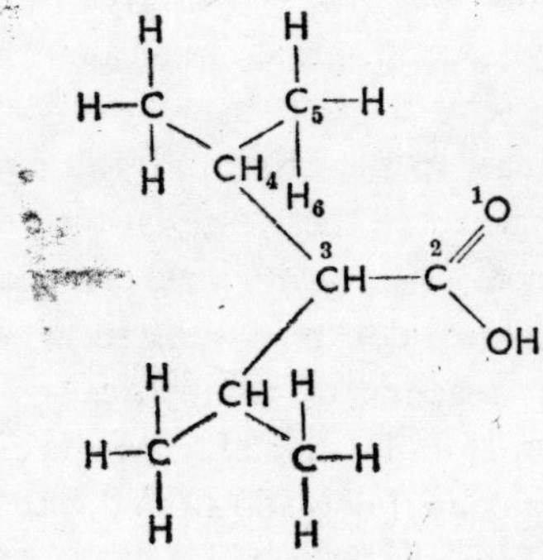

H H
H—C C_5—H
H CH_4 H_6
1 O
3 2
CH—C
OH
H CH H
H—C C—H
H H
V

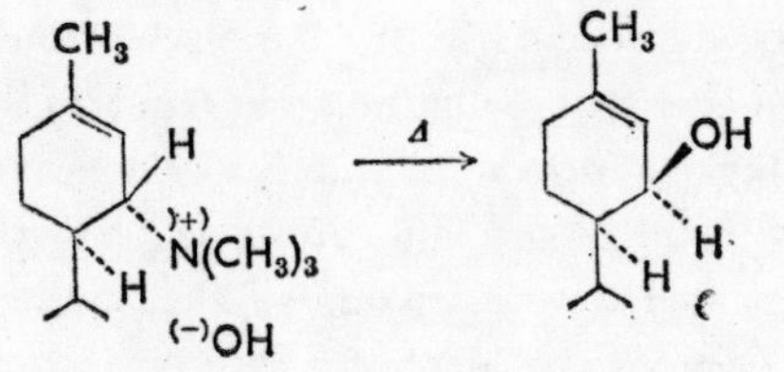

led Newman to the formulation of the empirical **rule of six**. According to this rule, the ease of esterification is decided by the number of atoms in position 6 from the oxygen atom of the carboxyl group. Aliphatic acids which have a large number of atoms in position 6 are difficult to esterify. For example, diisopropylacetic acid (*V*), which has 12 hydrogen atoms in position 6, cannot be esterified at all by usual methods. The Newman rule of six may also be applied to the saponification of amides, nitriles and esters of acids. In the esterification of a carboxyl group attached to an aromatic ring, the steric requirements of the substituents in *ortho*-position will be of paramount importance.

In the cyclohexane series, acids having their carboxyl group in an equatorial position are esterified more easily. The same is also true of the saponification of esters. For example the esters of *cis*-4-t-butylcyclohexanol are much more difficult to saponify than the respective *trans*-isomers. The saponification of *trans*-3-t-butylcyclohexanol esters with the carboxyl group in axial position is slower because of the same reason.

The alcohol moiety maintains its configuration during the esterification of organic acids and similarly in the course of the saponification of their esters, if it occurs by the bimolecular mechanism with the displacement of an acyl group. On the other hand, in the case of mineral acid esters, *i.e.* alkyl halides, sulphonic acid esters, and in some cases organic acid esters with tertiary and certain secondary alcohols, saponification is accompanied by the scission of the bond between the alkyl group and the oxygen of the alcoholic hydroxyl group, which leads to racemization of the alcohols.

The opening of epoxide rings may occur according to the S_N1 as well as S_N2 mechanism. The second one is much more usual and has been encountered in the course of the acid- as well as base-catalysed cleavage of epoxides and in the case of their reaction with amines, alkoxides *etc.* The configuration changes in all these cases at the carbon atom at which the epoxide ring is opened. *cis*-Epoxides yield compounds of *threo*-configuration, whereas *trans*-epoxides give substances with *erythro*-configuration. The hydrolytic reaction of symmetrically substituted *cis*-epoxides gives racemic vicinal diols, that of *trans*-epoxides yields the *meso*-form of diols. If the epoxide ring is substituted with a phenyl or another suitable unsaturated group, the epoxide ring may open by means of the

S_N1 mechanism and the products formed are not sterically uniform any more.

For the epoxides of compounds with a rigid structure (steroid epoxides) it is true that the entering substituent and the hydroxyl group formed occupy the axial position (the Fürst-Plattner rule). The epoxides of carbohydrates may appear in two conformations, which are able to convert one to the other. A rigid conformation is achieved only by adding another ring (*e.g.* by the synthesis of a benzylidene derivative). Methyl 2,3-anhydro-4,6-*O*-benzylidene-α-D-allopyranoside yields, on saponification, 84 % of methyl 4,6-*O*-benzylidene-α-D-altropyranoside and only 7 % of methyl 4,6-*O*-benzylidene-α-D-glucopyranoside.

Substitution Reactions with Cyclic Compounds

The reactivity of cyclic compounds is influenced by the size of the ring and its conformation, and is closely connected with the internal strain of these compounds. The internal strain, which Brown calls I-strain, includes the distortion of bond angles (Baeyer strain) as well as the nonbonded interactions present (Pitzer strain). In the course of the reaction accompanied by a change of the coordination number the strain increases or decreases, which affects the ease of substitution (see p. 119). The change of the coordination number from 4 to 3 accompanies the unimolecular substitution reactions S_N1 and S_H1; the change from 4 to 5 is involved in going to the transition state of bimolecular reactions (S_N2). The cycloalkyl halides and sulphonyl esters of cycloalkanols are, in the case of small (three- or four-membered) rings, extremely inert to all substitution reactions. The change of the coordination number from 4 to 3 in the instance of the S_N1 reactions and similarly from 4 to 5 in the case of S_N2 reactions is accompanied by the increase of internal strain, brought about mainly by the distortion of bond angles. With cyclopentane derivatives, the distortion of bond angles is insignificant and therefore the internal strain is influenced mainly by nonbonded interactions. The change of the coordination number from 4 to 3 and also from 4 to 5 is accompanied by a lessening of these interactions, and the reactivity is thus increased. Similar conclusions are also valid for the S_N1 reactions of compounds with medium rings. Substitution reactions in the case of six-membered and large rings occur, however, more slowly because the change of the coordination number is accompanied by an increase of internal strain. In the equilibrium state these cyclic compounds occupy conformations without steric strain; formation of a planar arrangement at the reacting carbon atom is not favoured as it disrupts the optimum energy arrangement (see p. 120). The changes of the coordination number of course refer to the carbon atom at which the substitution occurs.

When a group attached to a cyclohexane ring by means of an axial bond is substituted, in the transition state of the S_N2 reaction the original substituent starts to interact with the axial hydrogen atoms in positions 3 and 5. In the case of an equatorial substituent, these interactions affect the introduced particle Y and the original substituent is influenced by interactions with hydrogen atoms at the carbon atoms $C_{(2)}$ and $C_{(6)}$. Although steric interactions in the transition state are the same whether the substituent is in an axial or equatorial position, an axial substituent is substituted more easily because the activation energy of the reaction of the conformer with the axial substituent is smaller, thus facilitating substitution (see p. 121). S_N1 reactions require no special orientation of the breaking bond and therefore the difference in the reactivity of the axial and equatorial substituents is not large. Rigid systems, *e.g.* steroid, *etc.*, are suitable substances for the study of the reactivity of axial and equatorial substituents. In the course of the reaction of nitrous acid with amino derivatives of 5β-cholestane and 5α-cholestane, the equatorial amino groups quantitatively yield equatorial alcohols; the axial ones display a change of configuration accompanied by elimination.

Whilst S_N1-type reactions are the fastest in the field of medium rings, S_N2 reactions are very slow. This difference in reactivity is usually explained by the simultaneous action of the coplanar and collinear effects, the influence of which differs for the individual rings. The coplanar effect is connected with the formation of a planar arrangement at the carbon atom which is a member of the ring and acts in the same direction as the change of I-strain in the course of S_N1 reactions. This effect increases the reactivity of medium ring cycloalkyl derivatives with regard to the reactivity of similar acyclic derivatives. The collinear effect is connected with the collinear arrangement of the bonds in the transition state and increases its potential energy. Conversely, this effect reduces the reactivity of medium ring cycloalkyl derivatives.

Nucleophilic Substitution Reactions Occurring with Participation of Neighbouring Groups

The reactions of organic compounds are accelerated or retarded by the presence of atoms or groups of atoms in the vicinity of the reacting centre (1, 2). This effect is directly connected with the inductive, mesomeric and steric effects of such groups. If such a group is capable of forming a transient or pemanent bond with the carbon atom to which the reacting substituent is attached, and if this effect takes place in the step determining the reaction rate, then the reaction is greatly accelerated in comparison with a similar reaction where the effect is not encountered.

If a group adjacent to the reacting carbon atom takes part in the substitution reaction, a cyclic compound is formed first, which may be either a stable compound (epoxides, episulphides, epimines), or an unstable arrangement usually carrying a positive charge.* When the initial cyclic product is unstable, the structure of the reaction products depends on the place at which the transiently formed ring is opened in the successive reaction (see Diagram D 3). If group Z would not take part in the formation of the cyclic transition state, an S_N1 reaction would result in the formation of a classical ion and by combination with particle Y, the ion would yield a mixture of two diastereoisomeric products. The S_N2 reaction at the substituted carbon atom would cause an inversion of the configuration. Walden inversion usually accompanies the closing or opening of a three-membered ring. If both processes occur at the same carbon atom, we obtain a product with the original configuration. If we start from a substance (*VI*) in which the respective groups occupy a *threo*-configuration, we obtain a substitution product of

* In early British literature this cyclic ion was called a synartetic ion and the increase of the reaction rate as synartetic effect. The term "anchimeric assistance" is now commonly used.

threo-configuration (*VII*) as well; in the case of the *erythro*-configuration (*VIII*) the course of the reaction is similar and leads to compound *IX* (Diagram D 6).

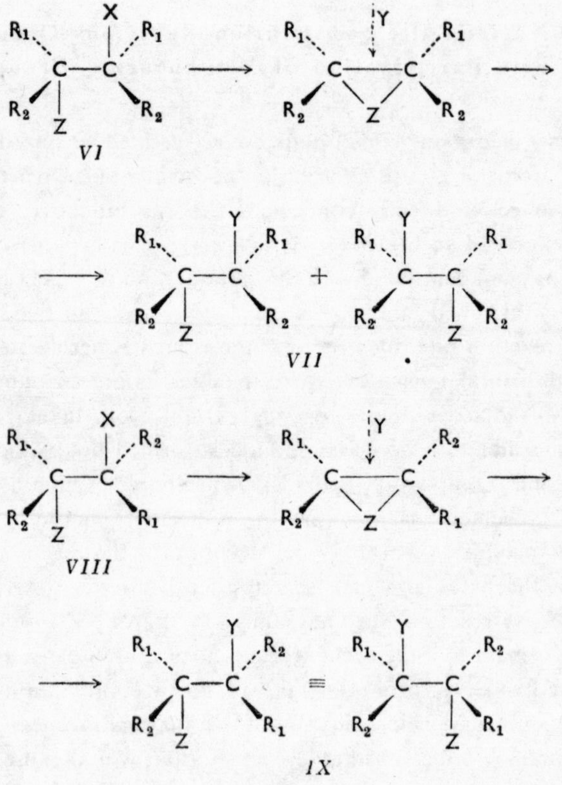

Diagram D 6

If the attacking particle Y is a solvent molecule, we again encounter a solvolysis reaction. This type of reaction is very important for the investigation of relative reaction rates and thus also for the quantitative evaluation of the influence of neighbouring groups on the course of substitution.

Halogen as Neighbouring Group

The reaction of diastereoisomeric 3-bromo-2-butanols with hydrobromic acid or phosphorus tribromide produces dibromo derivatives with the same configuration. The reaction course can be explained acceptably by the formation of a cyclic bromonium ion, identical with the postulated intermediate ion in the addition of bromine to an unsaturated compound. If we start with the optically active *threo*-isomer *X* we obtain racemic *threo*-2,3-dibromobutane (*XII*), the reaction proceeding through the symmetrical cyclic bromonium ion (*XI*). The *erythro*-isomer (*XIII*) yields the *meso*-form of *erythro*-2,3-dibromobutane (*XIV*); see Diagram D 7. An analogous course is taken by, for example, the reaction

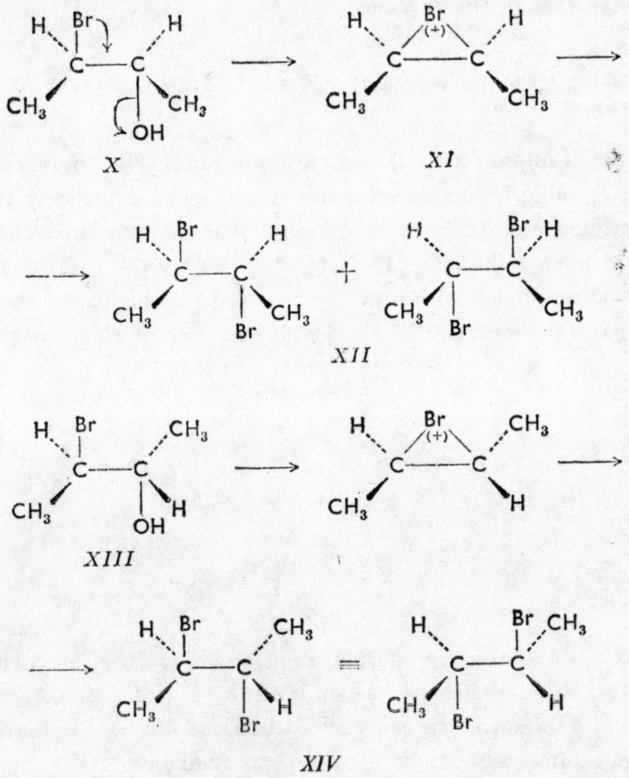

H Br H C—C CH_3 CH_3 OH → H Br (+) H C—C CH_3 CH_3 →

X *XI*

→ H Br H C—C CH_3 CH_3 Br + H Br H C—C CH_3 Br CH_3

XII

H Br CH_3 C—C CH_3 OH H → H Br (+) CH_3 C—C CH_3 H →

XIII

→ H Br CH_3 C—C CH_3 Br H ≡ H Br CH_3 C—C CH_3 Br H

XIV

Diagram D7

of both 2,3-dibromobutanes with silver acetate in acetic acid. An analogous course may also be expected of compounds containing chlorine or iodine; in some cases stable bromonium or iodonium salts have even been isolated. The example of the solvolytic reaction of *p*-bromobenzenesulphonates of *cis*- and *trans*-2-halogenocyclohexanols shows us how the character of the halogen influences ionization and thus also the reaction rate. The reaction of acetic acid with the *p*-bromobenzenesulphonate of *trans*-2-iodocyclohexanol, which undergoes solvolysis three million times as quickly as the analogous *cis*-isomer, is 1,600 times faster than the reaction of the unsubstituted *p*-bromobenzenesulphonate of cyclohexanol. On the other hand, the solvolysis of the *p*-bromobenzenesulphonate of *trans*-2-chlorocyclohexanol proceeds 2,500 times more slowly than that of the *cis*-isomer.

Neighbouring Groups Containing Atoms of Sulphur, Oxygen and Nitrogen

An atom of sulphur may take part in a number of substitution reactions either by forming isolatable sulphonium salts or by influencing the rate of a substitution reaction by its presence. For example, *trans*-2-chlorocyclohexyl phenyl sulphide (*XV*), in which both groups, S—C_6H_5 and Cl, may occupy the axial position, reacts at a rate higher by about five orders than the respective *cis*-isomer. In the case of rigid systems like

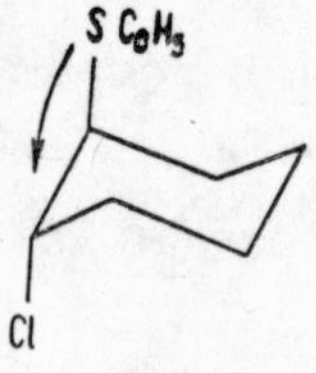

XV 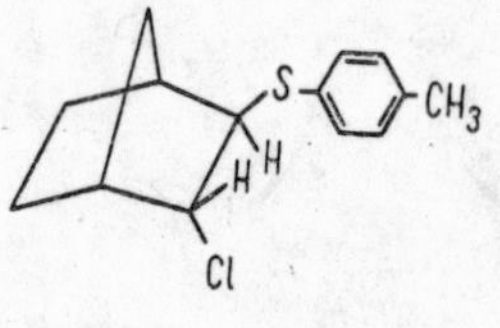 *XVI*

derivative *XVI*, the groups participating in the reaction are unable to occupy the required *anti*-periplanar position in either of the isomers, and therefore the solvolysis proceeds at the same slow rate in both cases. The substitution reactions of 1,2-aminochloroalkanes (*XVII*) are often accompanied by isomerization (see Diagram D 8).

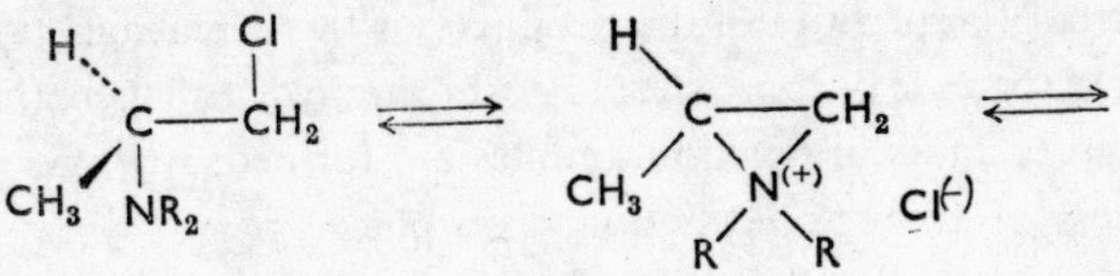

H Cl C—CH_2 CH_3 NR_2 ⇌ H C—CH_2 CH_3 $N^{(+)}$ R R $Cl^{(-)}$ ⇌

XVII

⇌ H Cl C—CH_2 CH_3 NR_2

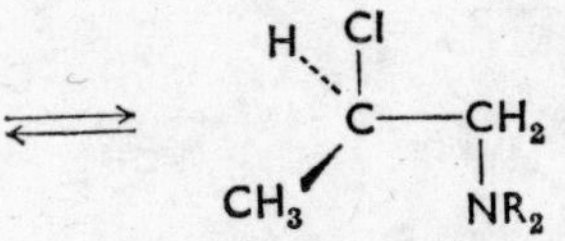

Diagram D 8

However, the configuration is retained in the course of the substitution reactions of chlorine in 3-chlorotropane (*XVIII*) because the conformation of the molecule permits the interaction of the tertiary nitrogen atom (see Diagram D 9). The synthesis of epoxides from halogenohydrins constitutes a preparatively important reaction which is affected by the oxygen atom. If we compare the reaction rates of ethylene chlorohydrin and ethyl chloride with sodium methoxide, we discover that it is 5,100 times larger in the first case. The reaction requires the halogen atom and the hydroxyl group to occupy an *anti*-periplanar arrangement. In the aliphatic series, we may prepare epoxides from both the *erythro*- and the *threo*-halogenohydrins because of the possible rotation around the C—C bond. For example, in the case of 3-bromo-2-butanol, we obtain the *trans*-epoxide *XX* from the *erythro*-halogenohydrin *XIX*, and the *cis*-epoxide *XXII* from the *threo*-isomer *XXI* (see Diagram D 10).

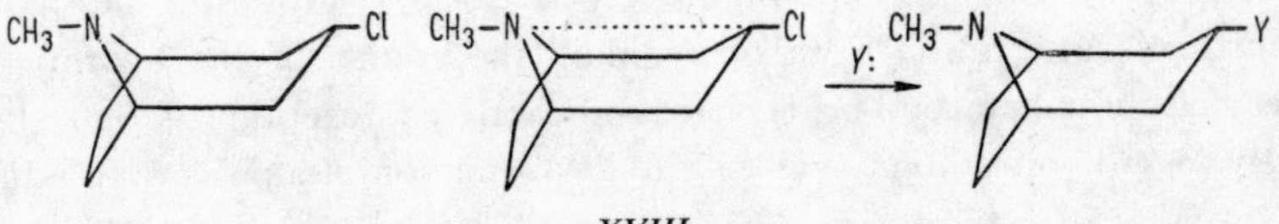

CH_3—N Cl CH_3—N Cl Y: → CH_3—N Y

XVIII

Diagram D 9

In the course of the alkaline saponification of halogenohydrins, we frequently isolate the products of epoxide ring opening. *trans*-Bromohydrins of the cyclohexane series in which both substituents are axial are converted to epoxides or the products of their cleavage, respectively. However, in the case of those steroid *cis*-bromohydrins where the bromine atom occupies an axial position ketones are formed, whereas if the bromine is equatorial ring contraction takes place.

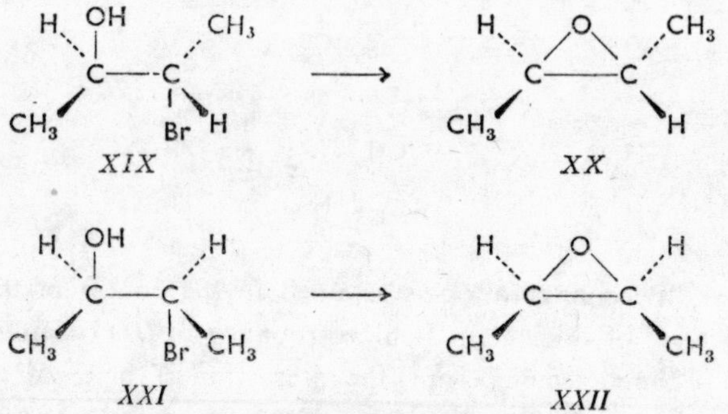

XIX → *XX*

XXI → *XXII*

Diagram D 10

The effect of the alkoxy group is not as large as the effect of the hydroxyl group. For example, both diastereoisomeric 2-bromo-3-methoxybutanes react with silver acetate in acetic acid, preserving their configuration. The toluene-*p*-sulphonate of *trans*-4-methoxycyclohexanol reacts at a six times greater rate than the *cis*-isomer and yields, besides the olefin, *trans*-4-methoxycyclohexyl acetate as the product of substitution.

Effect of Neighbouring Carboxyl Group

The first examples of the influence of a carboxyl group on the steric course of substitution were observed in the course of the hydrolytic cleavage of α-halogeno-acids. The acid-catalysed reaction of optically active α-bromopropionic acid leads to the formation of a mixture of both hydroxy acid enantiomers. The lactic acid with inverted configuration predominates in the mixture, as is required by the S_N2 reaction. However, when we use an alkaline medium we obtain lactic acid of the same configuration. The most probable explanation is that the intermediate formation of the unisolated α-lactone affects the reaction sequence (see Diagram D 11).

Similarly, the deamination of α-amino acids by the action of nitrous acid occurs with the preservation of configuration. An analogy may be found in the preparation of isolatable β-lactones from β-halogeno acids by the action of weak bases, occuring with an inversion of the configuration. Accordingly, the *threo*-isomer of β-halogenoacid (*XXIII*) yields lactone (*XXIV*) when treated with a base, whereas the *erythro*-isomer *XXV* yields hydroxy acid (*XXVI*) only.

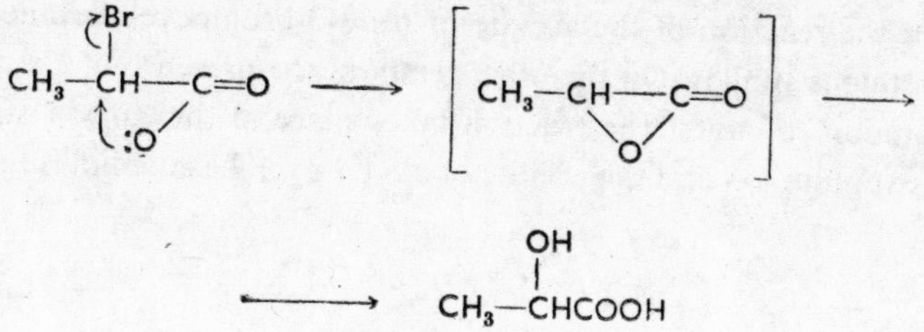

$$CH_3{-}CH(Br){-}C{=}O\ (:O^-) \longrightarrow \left[CH_3{-}CH{-}C{=}O \text{ (ring via O)} \right] \longrightarrow CH_3{-}CH(OH)COOH$$

Diagram D 11

The reaction mechanism of the opening of β-lactone rings depends to a considerable extent on the reaction conditions. If the reaction takes place at the carbon atom of the carbonyl group, we obtain a product which retains its configuration (the reaction occurring in acidic as well as alkaline media). On the other hand, the reaction at the β-carbon atom (in a neutral medium) leads to a product with inverse configuration:

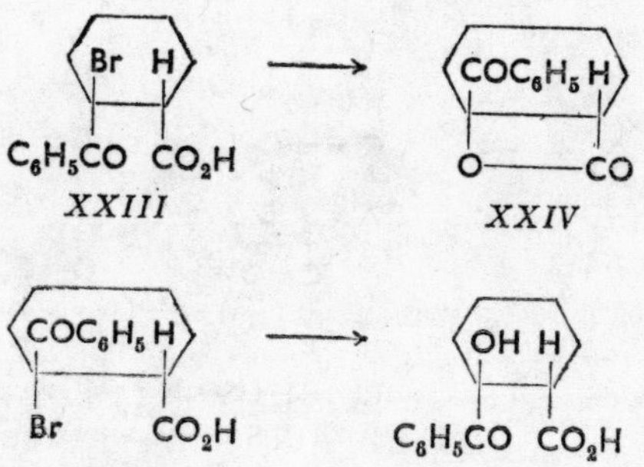

Br H / C_6H_5CO CO_2H (*XXIII*) $\longrightarrow$ COC_6H_5 H / O—CO (*XXIV*)

COC_6H_5 H / Br CO_2H (*XXV*) $\longrightarrow$ OH H / C_6H_5CO CO_2H (*XXVI*)

Effect of Neighbouring Ester and Amide Group

The stereochemical course of the reactions of this group depends very much on the experimental conditions. For example the reaction of *trans*-2-acetoxycyclohexyl *p*-bromobenzenesulphonate in the presence of sodium acetate proceeds through the acetoxonium ion stage and leads to the acetate of *trans*-1,2-cyclohexanediol.

The *cis*-isomer, in which the ester group cannot participate in the course of the solvolytic reaction, reacts at a 10^4 times slower rate. The course of the reaction of the acetate of *trans*-2-bromocyclohexanol with silver acetate is similar. On the other hand, in the presence of a stoichiometric amount of water the reaction takes place at the carbon atom of the acetoxonium ion and the acetate of *cis*-1,2-cyclohexanediol is formed.

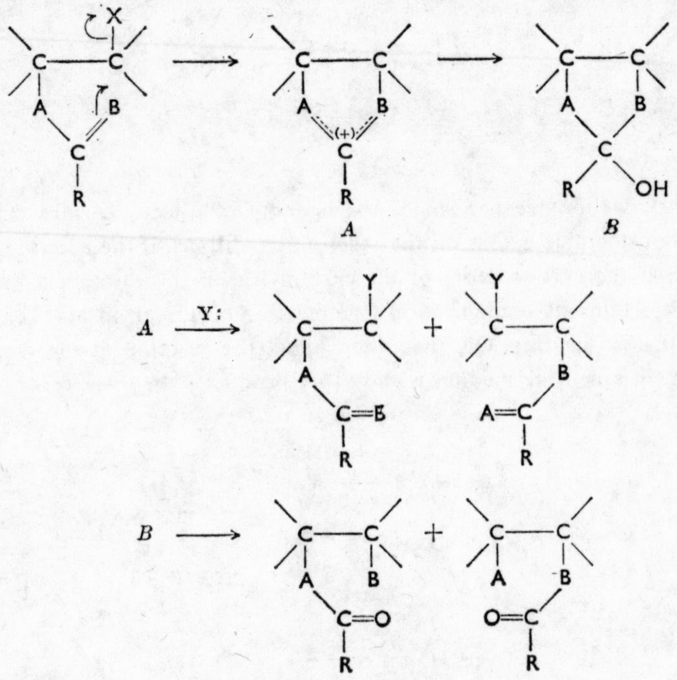

X = Cl, OH, OSO_2R
A = B = O, S, N
R = alkyl, aryl
Diagram D 12

In general we may say that the less nucleophilic reagents support the reaction at the carbon atom of the non-classical ion being formed and lead to a product with an inverse configuration at the substituted carbon atom. Conversely, strongly nucleophilic reagents yield products which retain their configuration. The reaction course af all these reactions is shown in Diagram D 12.

The amide group behaves analogously to the ester group. The toluene-*p*-sulphonyl ester of *trans*-2-benzamidocyclohexanol reacts at a 100 times greater rate than the *cis*-isomer and at a 200 times greater rate than the toluene-*p*-sulphonyl ester of *cis*-2-acetoxycyclohexanol. The effect of the benzamido group is thus even more marked than in the case of the ester group. The reaction of the *trans*-isomer with acetic acid in an anhydrous medium leads again to the acetate of the alcohol with inverse configuration, whereas in the presence of water the acetate with unchanged configuration was obtained.

Reactions of this type also include N → O migrations of acyl groups (see Diagram D 13). The *threo*-isomers of aliphatic amino alcohols (*e.g.* pseudoephedrine) react at a much greater rate than the *erythro*-isomers (*e.g.* ephedrine). The difference is caused by the fact that migration requires a *syn*-periplanar arrangement of the reacting groups in the transition state and such an arrangement is hindered by the nonbonded interactions of the *erythro*-isomer. Sicher *et al.* investigated this reaction

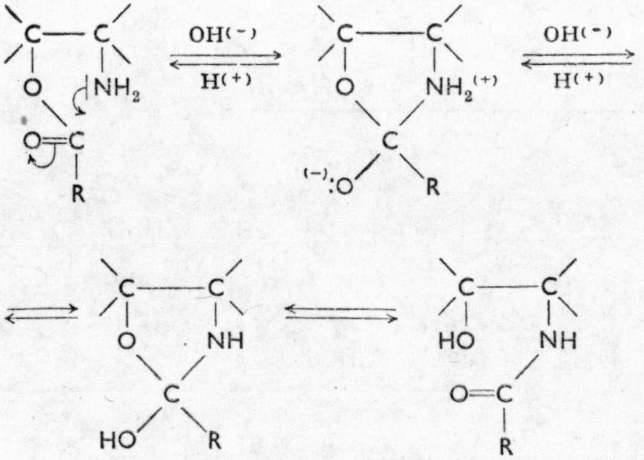

Diagram D 13

on the *N*-thiobenzoyl derivatives of epimeric 1,2-aminoalcohols and established that in this case *threo*-derivatives yield *trans*-Δ^2-oxazolines exclusively, and that *erythro*-isomers give mainly *trans*-Δ^2-thiazolines besides a small amount of *cis*-Δ^2-oxazolines.

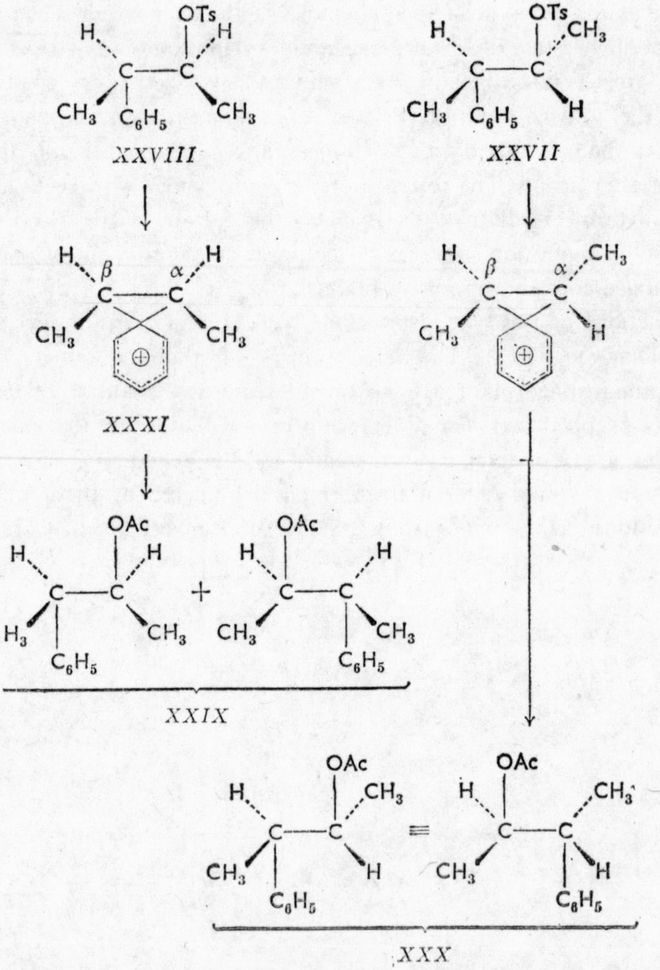

XXVIII *XXVII*

XXXI

XXIX

XXX

Diagram D 14

Effect of Neighbouring Phenyl Group

In all the cases stated until now, the groups present in the reacting molecule and influencing the course of the substitution had at least one free electron pair available. As the aromatic ring also has nucleophilic properties to a certain extent, we may expect that under certain conditions it will influence the substitution by the formation of a cyclic, so-called phenonium, ion. The first to prove the validity of this assumption was Cram in the course of the reaction of stereoisomeric toluene-*p*-sulphonates of 3-phenyl-2-butanols with acetic acid. When reacting optically active *erythro*-isomer (*XXVII*) or *threo*-isomer (*XXVIII*), he obtained in both cases 3-phenyl-2-butyl acetates of the same configuration as the original alcohol. The difference was only in the fact that the optically active *threo*-isomer yielded racemic product *XXIX* only, and the *erythro*-isomer yielded optically active product *XXX*. The absence of the *erythro*-isomer in the reaction products when we start from the *threo*-isomer and *vice versa* shows that in the course of the formation of the phenonium ion a change of the configuration at C_α takes place, and when it opens the configuration changes either at C_α or at C_β. The loss of optical activity in the case of the *threo*-isomer is caused by the formation of the absolutely symmetrical ion *XXXI* (see Diagram D 14).

Effect of Neighbouring Double Bond

For some time it has been known that the toluene-*p*-sulphonyl esters of steroid 3 β-hydroxyderivatives containing a double bond between the carbon atoms $C_{(5)}$ and $C_{(6)}$ react with nucleophiles, retaining the configuration at $C_{(3)}$. A difference in reactivity caused by a double bond has been observed for many bicyclic compounds. For example, of the toluene-*p*-sulphonyl esters of 7-hydroxybicyclo[2,2,1]hept-2-ene, the *anti*-isomer *XXXII* reacts at a 10^{11} times greater rate than the saturated compound and the *syn*-isomer *XXXIII* reacts at a 10^{7} times lesser rate than the *anti*-isomer, yielding mainly the rearranged product *XXXIV* (see Diagram D 15). The double bond influences the course of the reaction in all the cases referred to.

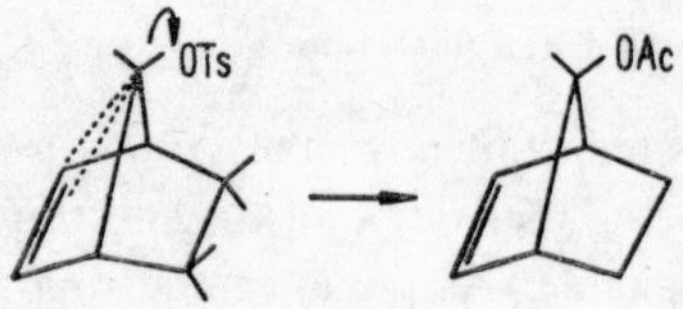

OTs OAc

XXXII

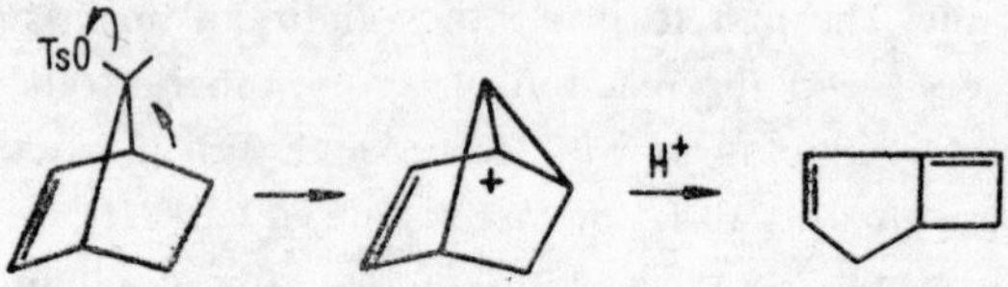

TsO H$^+$

XXXIII *XXXIV*

Diagram D 15

Effect of Neighbouring Hydrogen Atom and Hydrocarbon Group

Many organic reactions involve the migration of hydrogen atoms. By way of an example we may mention the reaction of the toluene-*p*-sulphonate of 2-butanol-[$1-{}^{14}C$] with acetic acid. The distribution of radioactivity in the reaction products testifies to a 1,2-shift of the hydrogen atom. The ethanolysis of *cis*-2-methylcyclohexyl toluene-*p*-sulphonate, in which the hydrogen can be in an *anti*-periplanar position to the substituent group and facilitate its ionization, proceeds at a 71 times greater rate than with the *trans*-isomer and leads to a substitution product of *cis*-configuration. Methyl shifts are common for instance in the course of the reactions of neopentyl derivatives.

The Stereochemical Course of Molecular Rearrangements

Many molecular rearrangements may be explained by the 1,2-shift mechanism, which is very similar to substitution reactions occurring with the participation of the groups vicinal to the reacting carbon atom. The difference is in the fact that the migrating group is a hydrocarbon group together with its bonding electrons. If the migrating group is attached *via* an asymmetric carbon atom, the latter preserves its configuration even in the final product. Conversely, a change of configuration takes place at the carbon atom to which the hydrocarbon group migrates. The stereochemical course may be explained satisfactorily by the assumption that the forming and breaking of bonds is a simultaneous process and that its energy is lowest if the migrating group replaces particle **X** from the opposite side.

The preparation of amines from acids and their derivatives forms a large and important group of reactions. The **Hofmann degradation** of acid amides as well as the **reactions of Curtius, Schmidt and Lossen** lead to amines of the same configuration (see Diagram D 16). Racemization also does not occur in the case of the degradation of acid amides displaying atropisomerism (*e.g. XXXIV*).

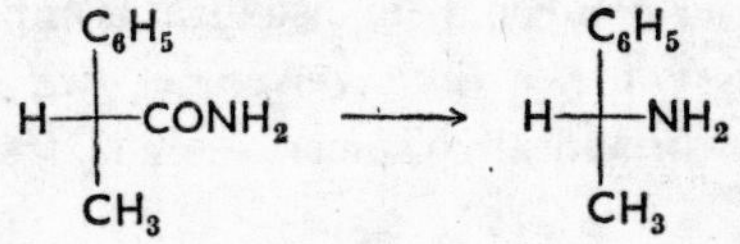

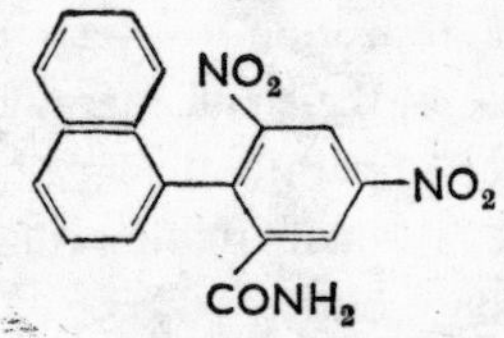

C_6H_5 / $H-C-CONH_2$ / CH_3 ⟶ C_6H_5 / $H-C-NH_2$ / CH_3

NO_2 / $-NO_2$ / $CONH_2$

XXXIV

Diagram D 16

The **Wolf rearrangement** is similar to the rearrangements referred to. The Wolf rearrangement forms a ketene from diazoketones and is the basis of the chain extension of acids according to Arndt and Eistert (see Diagram D 17). The **Beckmann rearrangement** of oximes (3), a further

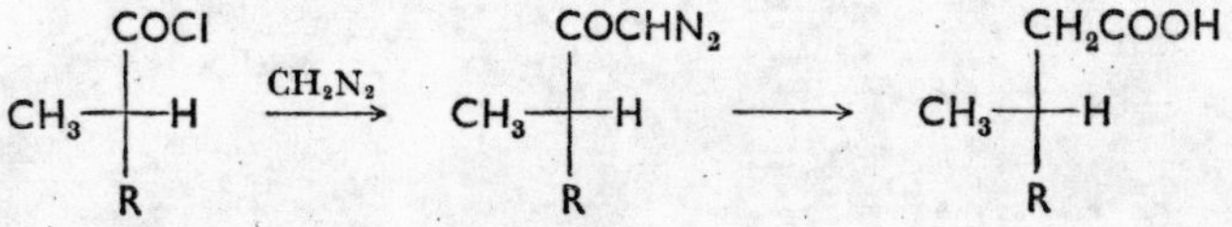

$COCl$ / CH_3-C-H / R $\xrightarrow{CH_2N_2}$ $COCHN_2$ / CH_3-C-H / R ⟶ CH_2COOH / CH_3-C-H / R

Diagram D 17

reaction of this type, occurs when the migrating group attacks the nitrogen atom from the opposite side to where the hydroxy group is located; if the group is chiral, it migrates without change of configuration (see Diagram 18).

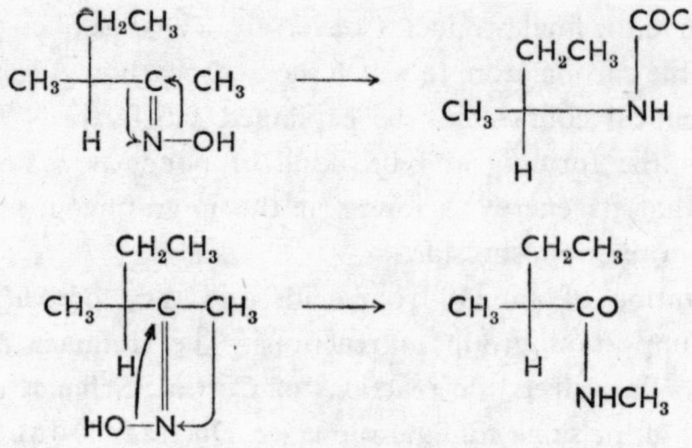

Diagram D 18

The transition between the *syn*-oxime and the *anti*-oxime is easy because of the small energy barrier between them; usually it is impossible to even separate them. Because of easy interconversion, we always isolate a small amount of the structurally isomeric amide from the reaction product.

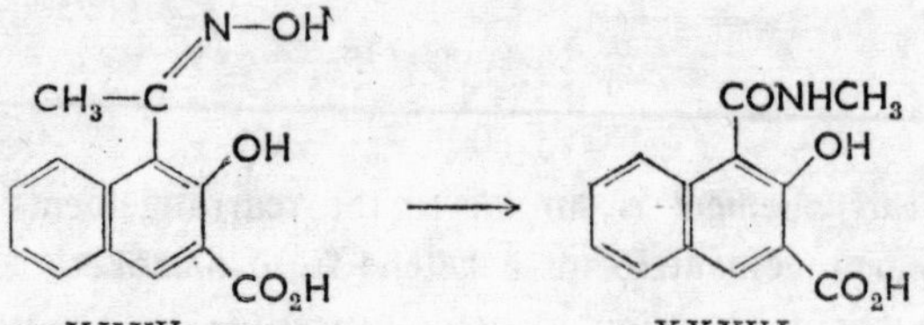

XXXV XXXVI

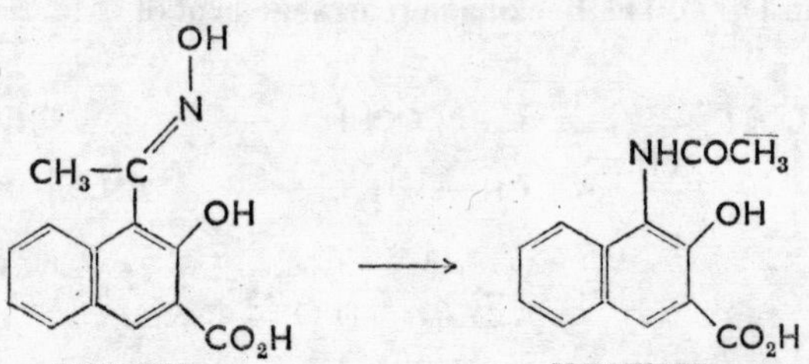

XXXVII XXXVIII

The acidic reagents used for the rearrangement act catalytically in the course of the conversion of the less stable oxime form into the more stable one (see p. 18).

For example, the rearrangement of the oxime of 1-acetyl-2-hydroxy-3-naphthoic acid (*XXXV*), which is resolvable into optical enantiomers, leads to the methylamide of 2-hydroxynaphthalene-1,3-dicarboxylic acid (*XXXVI*). The nonresolvable form *XXXVII* yields the 1-acetylamino derivative *XXXVIII*.

The reactions of ketones with peracids occur also with retention of the configuration (the **Baeyer–Villiger reaction**), as is shown in Diagram D 19. The reaction was used for the determination of the configura-

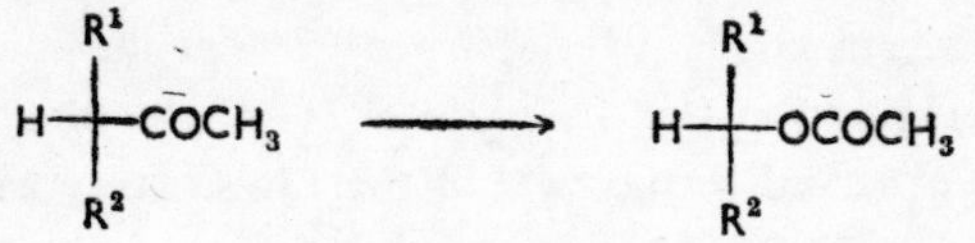

R^1 H—C—$COCH_3$ R^2 ⟶ R^1 H—C—$OCOCH_3$ R^2

Diagram D 19

tion of certain steroid compounds. In the case of the **pinacolone, retro-pinacolone** and **Wagner-Meerwein rearrangements** we encounter a similar reaction course. The differences in reactivity are connected with the structure of the starting material and with the energy difference of the respective transition states. Therefore, *cis*-1,2-dimethyl-1,2-cyclopentanediol is rearranged more easily than the corresponding *trans*-isomer. The reaction of 1,2-amino alcohols with nitrous acid is analogous to the pinacolone rearrangement as it is also connected with an inversion

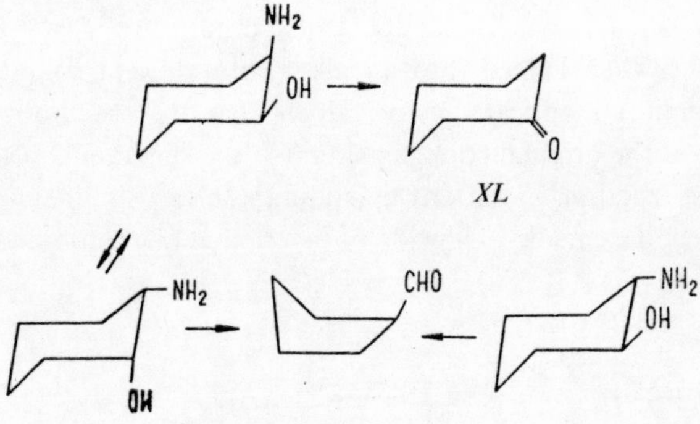

NH_2 OH ⟶ O

XL

NH_2 OH ⟶ CHO ⟵ NH_2 OH

XXXIX *XLI* *XLII*

Diagram D 20

of the configuration at the carbon atom to which the rearranging group migrates. A simple manifestation of the conformational effects in the course of the stated rearrangement is given by the reaction of diastereoisomeric 2-aminocyclohexanols with nitrous acid. The *cis*-isomer *XXXIX* yields cyclohexanone (*XL*) and 1-formylcyclopentane (*XLI*), whereas the *trans*-isomer *XLII* yields 1-formylcyclopentane only (4); see Diagram D 20. The majority of electrophilic rearrangements, in the course of which a group migrates without its bonding electrons, are characterized by a stereospecific course, even if to a lesser degree. For the **Steven's rearrangement** it was established (5) that the migrating group retains its configuration (see Diagram D 21). Retention of configuration was also observed in the case of the **Claisen rearrangement**.

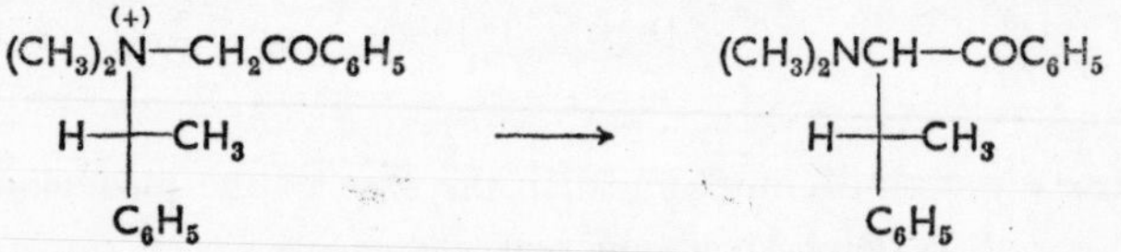

$(CH_3)_2\overset{(+)}{N}-CH_2COC_6H_5$ / $H-C-CH_3$ / C_6H_5 ⟶ $(CH_3)_2NCH-COC_6H_5$ / $H-C-CH_3$ / C_6H_5

Diagram D 21

Intramolecular Nucleophilic Substitution (S_Ni)

One of the most frequently used methods for the preparation of alkyl chlorides is the reaction of alcohols with thionyl chloride. It has been established that this reaction proceeds through the intermediate chlorosulphinate *XLIII*, which may be isolated in some cases and which decomposes when heated into an alkyl chloride and sulphur dioxide. If we submit an optically active alcohol to the reaction, we obtain a chloride of the original configuration (6) (see Diagram D 22). When we change the reaction conditions by adding a tertiary amine, or more frequently pyridine, an alkyl chloride of inverted configuration is produced.

CH_3, C_6H_5, H, O, S=O, Cl (*XLIII*) ⟶ CH_3, Cl, C_6H_5, H + SO_2

Diagram D 22

In the first case the thermal decomposition occurs by a cyclic (S_Ni) intramolecular mechanism; in the other case the added base supports dissociation to give chloride ions which subsequently react by an S_N2 mechanism. The decomposition of chloroformates has a similar course. An S_Ni reaction probably occurs even in the case of the reaction of hydrogen halides with alcohols at extremely low temperatures. (−)-α-Phenylethanol and anhydrous hydrogen bromide at −80 °C yield (−)-α-phenylethyl bromide of the same configuration. The higher the temperature, the greater the amount of the product with inverted configuration formed. The reactions of amines with nitrous acid or nitrosyl halides also sometimes occur by an S_Ni mechanism.

Intramolecular substitution has also been observed in the case of allyl compounds (S_Ni'). The reaction of thionyl chloride with optically active trans-α,γ-dimethylallylcarbinol (*XLIV*) yielded a chloride of opposite configuration (*XLV*) (see Diagram D 23).

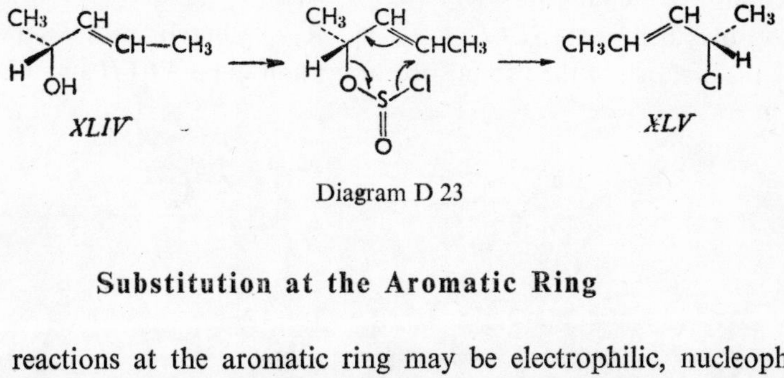

Diagram D 23

Substitution at the Aromatic Ring

The reactions at the aromatic ring may be electrophilic, nucleophilic or radical, as with aliphatic and alicyclic compounds. In the case of aromatic compounds, electrophilic substitution reactions are the most frequent, in contrast to aliphatic compounds. The ratio of the *ortho*:*para* isomers decreases with the increasing size of the alkyl group in the case of electrophilic substitution of alkylbenzenes. The greatest effect is usually encountered in the case of sulphonation, because the entering group is relatively the largest one. Nucleophilic substitution reactions occur with difficulty in the case of aromatic compounds, provided the substituted groups are not activated by groups attracting electrons. The S_N2 reaction is accompanied by the loss of aromatic character; in the transition state of the substitution we have to assume a tetrahedral arrangement of the bonds at the substituted carbon (see Diagram D 24). Substitution reactions occur much more easily with

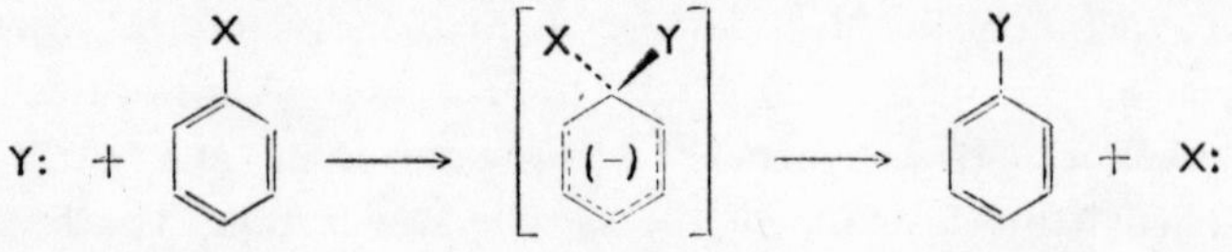

Diagram D 24

compounds where the reacting substituent, mostly a halogen or a nitro group, is activated by nitro groups in *ortho*- and *para*-positions. Here, for effective mesomeric stabilization of the transition state of the reaction, the oxygen atoms in the nitro group become coplanar with the ring, and this affects the course of the substitution. In the course of the reaction of 2,5-dinitro-*m*-xylene (*XLVI*, $R_1 = R_2 = CH_3$) or of 2,5-dinitro-t-butylbenzene (*XLVI*, $R_1 = H$, $R_2 =$ t-butyl) with ammonia, only the first one of the two possible transition states *XLVII* and *XLVIII*

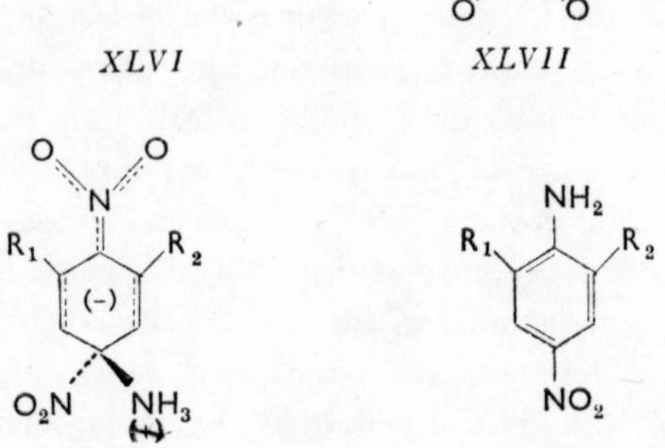

XLVI *XLVII*

XLVIII *XLIX*

Diagram D 25

will be encountered because only in that case can the nitro group be coplanar with the ring, thus unequivocally determining the structure of the substitution product (*XLIX*, see Diagram D 25).

Literature

1. Streitwieser A.: Chem. Revs. *56*, 675 (1956).
2. Lwovski W.: Angew. Chem. *70*, 483 (1958).
3. Blatt A. H.: Chem. Revs. *12*, 20 (1933).
4. Mc Casland G. E.: J. Am. Chem. Soc. *75*, 2293 (1951).
5. Brewster J. H., Klyne M. W.: J. Am. Chem. Soc. *74*, 5179 (1952).
6. Boozer C. E., Lewis E. S.: J. Am. Chem. Soc. *75*, 3182 (1953).

E. STEREOCHEMICAL COURSE OF ELIMINATION REACTIONS

General Comments Concerning Mechanism

Unsaturated compounds are generally prepared by elimination reactions, in which an organic molecule loses an atom or group of atoms attached to one carbon (C_α), and another atom (usually hydrogen) attached to the vicinal carbon (C_β). Such reactions are called β-eliminations. Water is eliminated from alcohols (X = OH, Y = H) in the case of dehydration, a hydrogen halide molecule (X = Cl, Br, I, Y = H) in the case of dehydrohalogenation, a halogen molecule (X – Y = Cl, Br, I) in the case of dehalogenation and a hydrogen molecule (X = = Y = H) in the case of dehydrogenation. The reactions referred to are supplemented by the cleavage of quaternary ammonium bases and sulphonium bases and by a number of other reactions of varying preparative importance. Generally speaking, elimination reactions may occur by radical as well as ionic mechanisms, the latter being more usual. Elimination reactions frequently accompany substitution reactions; it often depends on the structure of the substance and on the experimental conditions which of them will prevail. Ingold proposed that elimination reactions should be classified according to their molecularity as unimolecular (E_1 and E_1cb) and bimolecular (E_2).

Unimolecular elimination (E_1) is analogous to unimolecular nucleophilic substitution (S_Nl), and proceeds *via* a carbonium ion, the rate of formation of which determines the rate of the overall reaction. Then an electrophilic particle, usually a proton, is rapidly expelled (Diagram E 1). What has been said in connection with substitution reactions is also true of the structure of compounds displaying a tendency towards unimolecular elimination. The diagram seems to indicate that the character of particle X effects only the formation of the carbonium ion and not its subsequent changes. However, many cases have recently been encountered, when the eliminated group X considerably influences the result of the reaction. Thus for example in the solvolysis of 2-substituted 2-phenylbutanes the ratio of 2-phenyl-2-butene to 2-phenyl-1-butene decreases with the increasing basicity of the eliminated particle X.

$$\mathrm{H{-}C_{\beta}{-}C_{\alpha}{-}X} \xrightarrow[\text{(slow)}]{} \mathrm{H{-}C{-}C^{(+)}} + \mathrm{X^{(-)}}$$

$$\downarrow \; \mathrm{Y^{(-)}} \text{ (rapid)}$$

$$\underset{\mathrm{S_N1}}{\mathrm{H{-}C{-}C{-}Y}} \qquad \underset{\mathrm{E_1}}{\mathrm{>C{=}C<}} + \mathrm{HY}$$

Diagram E 1

It is assumed that anion $X^{(-)}$ remains in the vicinity of the carbonium ion after the rupture of the covalent bond and the ion-pair formed has different properties for different X. With increasing basicity of particle X, the ease of the transfer of a proton from the carbonium ion to this group increases, and results in augmenting the 1-alkene portion.

If the elimination is completed before complete dissociation of the ion-pair, then so-called *anti*-elimination occurs, even in the case of E_1 elimination. *anti*-Elimination has marked stereoelectronic requirements and requires the molecule to occupy a conformation in the transition state of elimination in which the eliminated groups (H and X in our case) would be in *anti*-periplanar position. The transition state is thus analogous to the transition state of E_2 reactions (see p. 168).

According to the **Saytzeff rule**, the thermodynamically more stable, most alkylated olefin usually predominates in the mixture of olefins formed. For example, the solvolytic reaction of t-pentyl bromide in 80 % ethanol yields, besides 60 % of substitution products, 32 % of 2-methyl-2-butene and only 8 % of 2-methyl-1-butene.

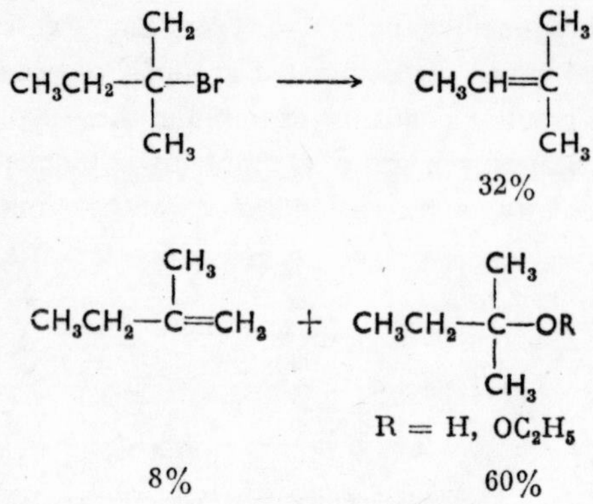

$$CH_3CH_2-\underset{CH_3}{\overset{CH_2}{C}}-Br \longrightarrow CH_3CH=\underset{CH_3}{\overset{CH_3}{C}}$$

32%

$$CH_3CH_2-\overset{CH_3}{C}=CH_2 + CH_3CH_2-\underset{CH_3}{\overset{CH_3}{C}}-OR$$

R = H, OC_2H_5

8% 60%

E_1cb reactions occur by a two-stage mechanism (Diagram E 2); the elimination of the proton is rapid and decomposition of the carbanion is the step determining the overall rate of the reaction:

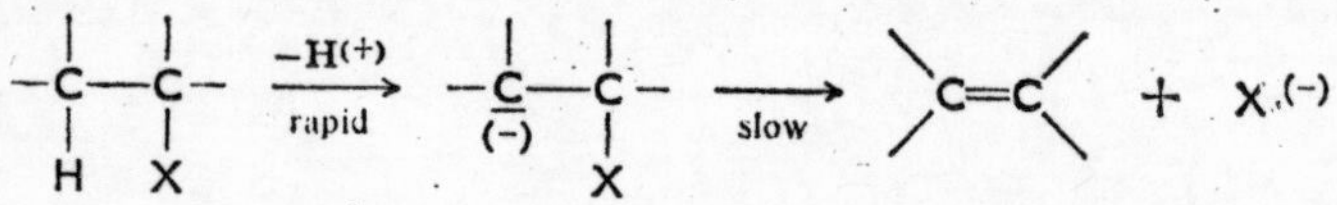

$$-\underset{H}{\overset{|}{C}}-\underset{X}{\overset{|}{C}}- \xrightarrow[\text{rapid}]{-H^{(+)}} -\underset{(-)}{\overset{|}{C}}-\underset{X}{\overset{|}{C}}- \xrightarrow[\text{slow}]{} \rangle C=C\langle + X^{(-)}$$

Diagram E 2

This type of mechanism is favoured especially if the β-hydrogen is made more acidic by the attached X groups. E_2 elimination reactions generally require an *anti*-periplanar arrangement of the ruptured bonds in the transition state. The basic reagent attacks the substituent at C_β and the leaving group is simultaneously split off; this process is facilitated by the stated

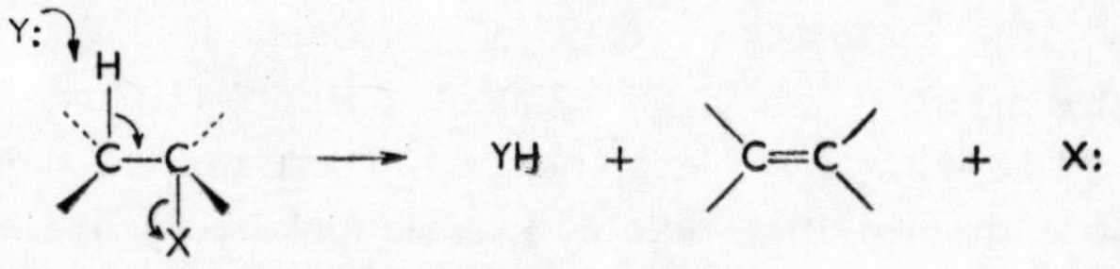

Y: H C—C X ⟶ YH + $\rangle C=C\langle$ + X:

Diagram E 3

partial conformation (Diagram E 3). If the stereoelectronic requirement is not fulfilled or if the achievement of the *anti*-periplanar state is connected with a substantial increase in nonbonded interactions, the E_2 elimination reaction proceeds sluggishly or not at all. Since elimination reactions are sensitive to polar factors, unequivocal proof of the influence of steric effects on the attainment of an *anti*-periplanar arrangement was obtained purely by study of reactivity of individual stereoisomeric pairs, where the influences other than steric effects are equal. The stereochemical course of bimolecular substitution reactions (S_N2) and elimination reactions (E_2) differs from the unimolecular reactions (S_N1) and (E_1) in the transition state of the reaction.

syn-Eliminations, which have a cyclic mechanism, represent a special case.

Mutual Dependence of Substitution and Elimination Reactions

The ratio in which substitution and elimination reactions proceed side by side is easily assessable if the reaction is unimolecular, because in this case the ratio S_N1/E_1 is almost independent of the character of the leaving group and practically also of the concentration and type of base in the solution. The generally valid rules governing the S_N/E ratio may be summarized as follows.

1. The introduction of substituents at C_α and C_β suppresses the substitution reaction considerably and conversely supports the elimination reaction.

2. Weak bases, which are however of markedly nucleophilic character ($C_6H_5S^-$, N_3^-, CN^-), facilitate substitution whereas strong bases (HO^-, RO^-), especially in high concentration, support elimination.

3. The higher the polarity of the solvent, the more the reactions S_N1 and E_1 are facilitated and, on the other hand, the lower its polarity, the more the S_N2 and E_2 reactions are supported.

Orientation During Elimination Reactions

In order to determine the probable structure of the predominant olefin in the course of E_2 elimination reactions we may apply two empirical rules. Dehydration and dehydrohalogenation are governed by the **Saytzeff rule**, according to which the most alkylated olefin is formed. In the course of cleavage of tetra-alkylammonium salts and trialkylsulphonium salts, however, the least alkylated olefin predominates in the reaction mixture according to the **Hofmann rule**. The Saytzeff rule is usually explained in terms of hyperconjugation. The electron pairs of the C—H bonds are more mobile than the electrons of C—C bonds. Therefore, in some cases we may encounter conjugation between the C—H bond electrons and the π-electrons of the double or triple bonds. The effect is the more pronounced, the larger the number of C—H bonds in the vicinity of the multiple bond. Consequently, it decreases in the order $CH_3C{=}C{-} > -CH_2C{=}C{-}CH{-}C{=}C{-}$. For dehydration and dehalogenation it is then true that the larger the number of C—H bonds attached in the vicinity of a double bond being formed, the more readily does elimination takes place.

$$\begin{array}{c} \text{H} \quad\;\; \text{Br} \;\; \text{H} \\ \text{H—C—CH—CH—C—H} \longrightarrow \text{HBr} + \\ \text{H} \;\; \text{H} \qquad\quad \text{H} \end{array}$$

$$+ \underset{71\,\%}{CH_3CH{=}CH{-}CH_3} + \underset{29\,\%}{CH_3CH_2CH{=}CH_2}$$

The reactions governed by the Hofmann rule are, on the other hand, influenced by the acidity of the hydrogen C_β atoms, which are eliminated. This depends on the number and character of the substituents attached to C_β, which facilitate the elimination of a proton by means of their inductive effect.

$$\underset{^{(+)}N(CH_3)_3OH^{(-)}}{CH_3{-}CH{-}CH_2{-}\text{alkyl}} \longrightarrow CH_2{=}CH{-}CH_2{-}\text{alkyl} + N(CH_3)_3$$

The stated explanation, based purely on electronic factors, is supported mainly by Ingold's school. For the sake of completeness we also include the opinion of Brown (1), who in contrast considers mainly steric effects, due not only to the structure of the molecule but also to the attacking reagent. The molecule may occupy a larger or smaller number of conformations, in which the eliminated atoms or groups of atoms are in *anti*-periplanar position. Of all the possible conformations complying with this condition, the most facile reaction will proceed *via* that conformation with the fewest nonbonded interactions. The elimination reactions of 2-substituted pentane may serve as an example. In the transition state of the elimination we may have the conformations illustrated by Formulas *I–III*. A 1-alkene is formed in the first case according to the Hofmann rule;

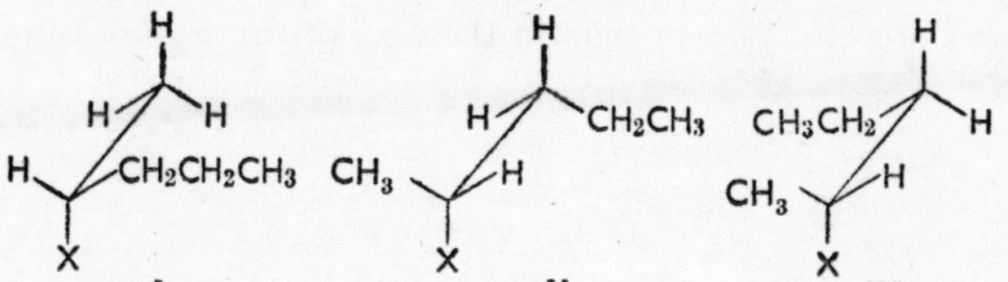

I *II* *III*

conformations *II* and *III* lead to 2-alkenes. Of the latter two, conformation *II* leading to *trans*-2-alkene is preferred on account of smaller steric interactions. The ratio of 1-alkene to 2-alkene depends on the character of the leaving group. Evidently, the larger the leaving group, the more will conformation *I* be preferred, because here the steric interactions between

$$Br < I < OSO_2C_7H_7 < \overset{(+)}{S}(CH_3)_2 < SO_2CH_3 < \overset{(+)}{N}(CH_3)_3$$

the leaving group and the chain alkyls are the fewest. The percentage of 1-alkene in the products will therefore grow. The ratio of 1-alkene to 2-alkene depends furthermore on the structure of the substrate and on the bulk of the attacking base. In both cases, the effect of the conformations with the smaller steric interactions and the different ease of access to the eliminated β-hydrogen will influence the ratio of isomeric olefins.

Dehydration and Dehydrohalogenation

The dehydration of alcohols and the dehydrohalogenation of alkyl halides occur by similar reaction mechanisms. Alcohols are usually dehydrated by acids or other electrophilic agents but β-hydroxycarbonyl compounds (aldols, β-hydroxy acids and their derivatives) are also dehydrated in an alkaline medium. On the other hand, alkyl halides eliminate hydrogen halides by the action of strong bases, in the course of which substitution reactions are suppressed to a minimum. Both reactions are markedly stereospecific. For example, in the course of the dehydrohalogenation of the two diastereoisomeric stilbene dibromides, the *erythro*-isomer *IV* yields *cis*-bromostilbene (*V*) and the *threo*-isomer *VI* yields *trans*-bromostilbene (*VII*). In the second case the elimination is facilitated because in the transition state of the reaction the steric interactions are much fewer than in the case of the *erythro*-isomer.

IV → *V*

VI → *VII*

Hydrogen iodide is generally eliminated from alkyl halides more easily than hydrogen bromide and hydrogen chloride. The condition of *anti*-periplanar arrangement of eliminated groups may always be fulfilled for non-cyclic compounds because of the possible rotation around the C_α—C_β bond. Conversely, in the case of cyclic compounds with a less than nine-membered ring, rotation and thus also the achievement of coplanarity is frequently prevented. Although in the case of *trans*-1,2-disubstituted cyclohexane derivatives both substituents usually are in the favoured equatorial position, they may, as a rule, pass into axial positions fulfilling all the conditions for bimolecular elimination. On the other hand, in the case of *cis*-1,2-disubstituted cyclohexane derivatives, one eliminated substituent always has to be in an equatorial position and the second one in an axial position. Without the introduction of considerable strain it is impossible to achieve an *anti*-periplanar arrangement. An illustrative example is given by the elimination of hydrogen chloride from menthyl chloride (*VIII*) and neomenthyl chloride (*IX*). The action of sodium methoxide produces, in the first case, only one olefin (2-menthene, *X*) but the reaction is much slower than that of neomenthyl chloride because the transition state of the reaction requires

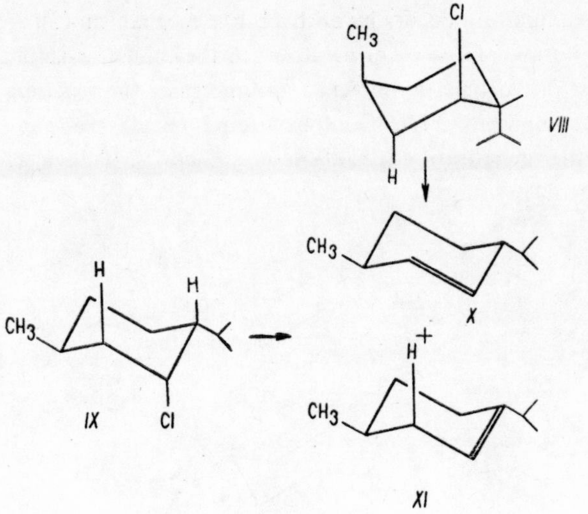

Diagram E 4

the chlorine to occupy an axial position, thus forcing the methyl and isopropyl groups into the unfavourable axial positions. On the other hand, if the chlorine atom of neomenthyl chloride is in an axial position, the methyl and isopropyl groups are in equatorial positions. The reaction produces a mixture of olefins, because the molecule contains two hydrogen atoms complying with the conditions of an E_2 reaction. 3-Menthene (*XI*) predominates in the mixture according to the Saytzeff rule (see Diagram E 4). The flexibility of the ring is further reduced in the case of bicyclic compounds. For example, in the case of dichlorodibenzo-[2,2,2]octadiene (*XII*), where the conditions of bimolecular elimination are not fulfilled, neither by the *trans*- nor by the *cis*-derivative, hydrogen chloride elimination proceeds extremely slowly. We have to assume

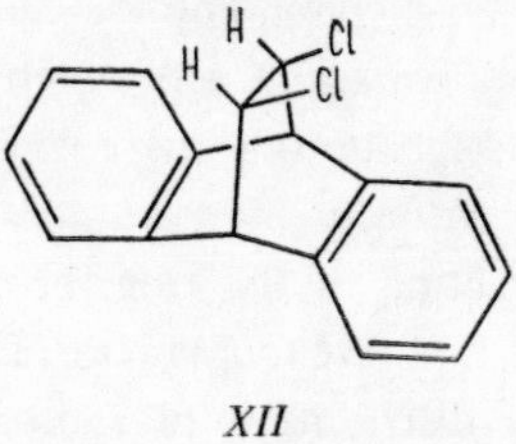

XII 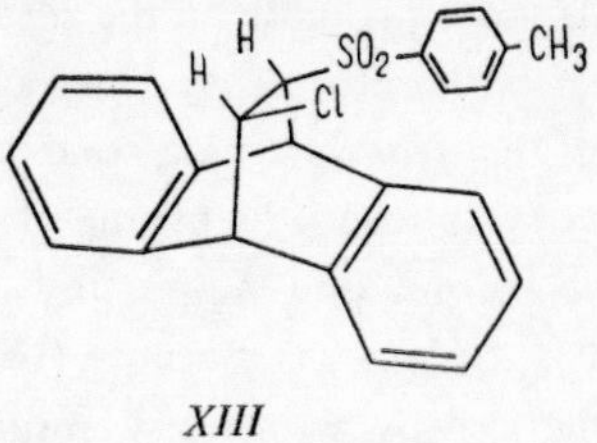 *XIII*

that the elimination occurs by an E_1cb-like mechanism (2). Accordingly, the substitution of one chlorine atom for the toluene-*p*-sulphonyl group makes the *β*-hydrogen more acidic and increases the reaction rate in the case of compound *XIII*. Conformational effects play an important role in the dehydration of diastereoisomeric 2-phenylcyclohexanols.

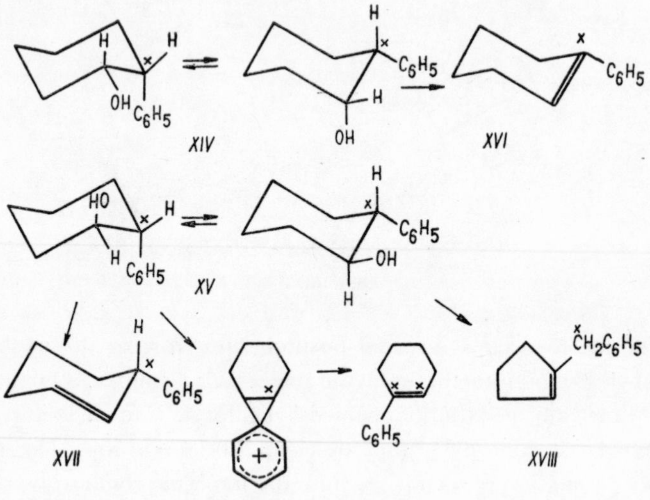

Diagram E 5

Dehydration of *cis*-2-phenylcyclohexanol-[2-^{14}C] (*XIV*) yielded 88 % of 1-phenylcyclohexene (*XVI*), and the *trans*-isomer *XV* yielded 21 % of 1-phenylcyclohexene (*XVI*), 9 % of 3-phenylcyclohexene (*XVII*) and almost 50 % of a mixture of 1-benzylcyclopentene (*XVIII*) with benzylidenecyclopentane. The conformation with the equatorial phenyl group is more probable in the case of the *cis*-alcohol, and the elimination of water is therefore easy. Two chair conformations are possible in the case of the *trans*-alcohol; one of them facilitating the formation of 1-phenylcyclohexene by means of a reaction occurring *via* a non-classical ion, the second one supporting a contraction of the ring, because the OH—$C_{(1)}$—$C_{(2)}$—$C_{(3)}$ atoms (Diagram E 5) are coplanar. (The positions marked by an asterisk indicate the distribution of radioactivity.)

The action of sodium methoxide on the diastereoisomeric toluene-*p*--sulphonates of 2-methylcyclohexanols cannot bring about the formation of a non-classical ion. The *cis*-isomer yields 36 % of 3-methylcyclohexene and 64 % of 2-methylcyclohexene, in accordance with the Saytzeff rule.

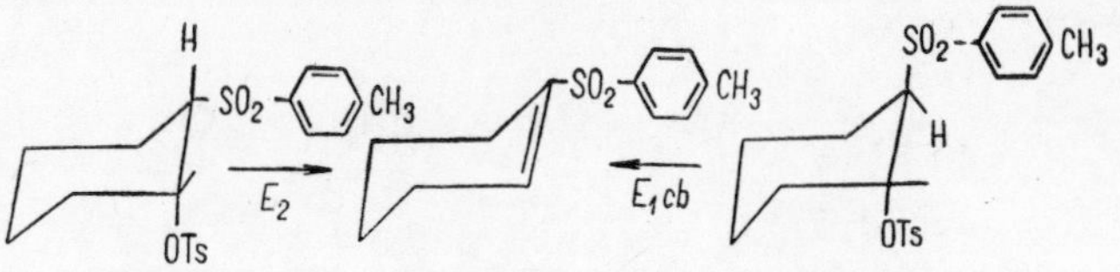

H
SO2 CH3 SO2 CH3 SO2 CH3
H
E2 E1cb
OTs OTs
XIX XXI XX

Diagram E 6

Only 3-methylcyclohexene has been isolated from the reaction products of the *trans*-isomer. If the methyl group is replaced by a toluene-*p*-sulphonyl group, elimination in the case of *cis*-isomer *XIX* occurs normally by the E_2 mechanism, but in the case of *trans*-isomer *XX* by the E_1cb-like mechanism. 1-Toluene-*p*-sulphonylcyclohexene (*XXI*) is the reaction product in both cases, as is shown by Diagram E 6.

Dehalogenation

Bimolecular dehalogenation by means of bivalent metals in a heterogeneous medium is in general characterized by a lower stereospecificity than dehalogenation in a homogeneous medium. The action of magnesium in tetrahydrofuran or of zinc in an aqueous medium produces up to 95 % of *trans*-2-butene from *meso*-2,3-dibromobutane (*XXII*); *cis*-2-butene has been obtained from the racemate *XXIII* (Diagram E 7). Similarly, dibromofumaric acid is dehalogenated at a three times greater rate than dibromomaleic acid, because in the first case the steric interactions in the transition state are fewer. On the other hand, dehalogenation by means of univalent metals usually occurs non-stereospecifically, which testifies to the radical mechanism of the reaction. Sodium in liquid ammonia reacts with *erythro*- and *threo*-2,3-dibromobutanes and produces a mixture of *cis*- and *trans*-2-butenes in a 1 : 1 ratio. Much lower stereospecificity is encountered if a group attracting electrons is

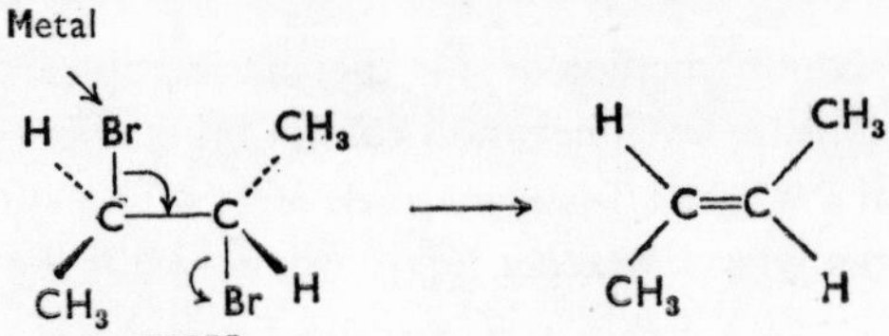

XXII

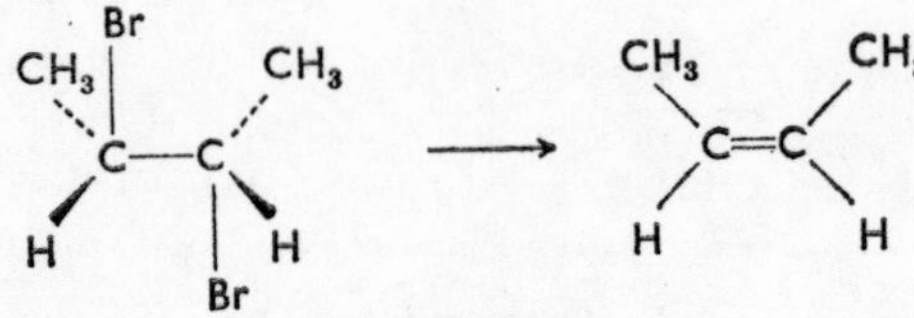

XXIII

Diagram E 7

attached to the carbon of the double bond being formed. For example, both diastereoisomeric 1,2-diphenyl-1,2-dibromoethanes yielded *trans*-stilbene only. The reactivity of vicinal dibromo derivatives is always higher than that of the dichloro derivatives. Therefore, alkali metal iodides may be used for their dehalogenation. The reaction is especially suited for kinetic studies as it occurs in a homogeneous medium. The iodide ion here takes over the role of the nucleophilic particle and attacks the bromine atom to give Br and at the same time the bromide ion is eliminated from the vicinal centre (Diagram E 8). Aliphatic *erythro*-1,2-dihalogeno derivatives (*XXIV*) eliminate bromine, as a result of smaller steric interactions in the transition state, more readily than do the *threo*-isomers. The significant difference in the ease of dehalogenation of compounds of the $R-CHBr-CH_2Br$ type compared to the $R-CHBr-CHBr-R$ type with the help of iodine is explained by the fact that in the former case substitution takes place first at the easily

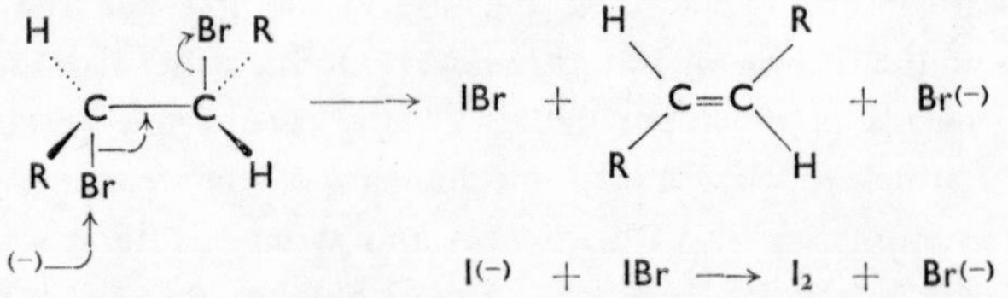

$$I^{(-)} + IBr \longrightarrow I_2 + Br^{(-)}$$

XXIV

Diagram E 8

accessible primary carbon and the iodine atom in the bromoiodo derivative formed is more susceptible to a nucleophilic attack than the bromine atom:

$$\mathrm{R{-}CHBr{-}CH_2Br} \xrightarrow[S_N2]{I^-} \mathrm{R{-}CHBr{-}CH_2I} \xrightarrow[E_2]{I^-} \mathrm{Br^-} + \mathrm{R{-}CH{=}CH_2} + \mathrm{I_2}$$

With cyclohexane derivatives, *trans*-1,2-dibromocyclohexane reacts at a 11.5 times greater rate than the analogous *cis*-isomer. In the case of the *cis*-isomer, where the steric conditions for an E_2 reaction are not fulfilled, it is probable that the elimination proper is preceded by a substitution of the axial bromine atom by an iodide ion, thus forming a *trans*-bromoiodo derivative, the elimination of which is relatively easy.

A special case of β-elimination is the dehalogenation-decarboxylation of β-halogenoacids to olefins (3). The reaction occurs stereospecifically in less polar solvents (ethanol); on the other hand, in water a mixture of olefins is formed in which the thermodynamically more stable *trans*-isomer predominates (probably due to the primary ionization of the C—Br bond). For example, the addition of bromine to crotonic acid (*XXV*) yields *erythro*-2,3-dibromobutyric acid (*XXVI*), from which we may obtain *cis*-propenyl bromide by the action of sodium carbonate (see Diagram E 9). The remaining reactions of this type will be discussed in the context of addition reactions (p. 188).

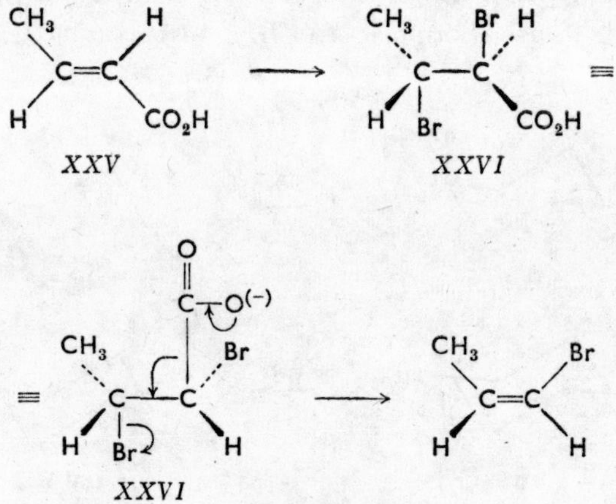

XXV → *XXVI* ≡

≡ *XXVI* → $\mathrm{CH_3(H)C{=}C(Br)H}$

Diagram E 9

The Stereochemical Course of Cleavage of Quaternary Ammonium Bases

The thermal cleavage of quaternary ammonium bases gives a tertiary amine, an olefin and water; only in occasional cases is an alcohol also isolated (4).

As a rule, the cleavage has a synchronous bimolecular E_2 mechanism. By attacking the hydrogen attached in the β-position to the trimethylammonium group with a base, a simultaneous scission of the C—H and $C—N^{(+)}(CH_3)_3$ bonds takes place, and a double bond is formed. Such a condition may be satisfied for aliphatic compounds in which at least one hydrogen may occupy an *anti*-periplanar position with respect to the trimethylammonium group. However, if this requirement is not fulfilled or if its achievement is connected with an increase of nonbonded interactions, the elimination reaction is either sluggish or does not take place at all. For instance, the pyrolytic decomposition of steroid trimethylammonium bases produces olefins in good yield only if the trimethylammonium group is in the axial position. In these rigid systems, if the trimethylammonium group is in the equatorial position, an *anti*-periplanar arrangement of the ruptured bonds is very difficult to achieve. Under conditions usually favourable for the elimination reaction, olefins are formed in only small yields and the formation of substitution products predominates. During the elimination reaction of *cis*- (Formula *XXVII*) and *trans*- (Formula *XXVIII*) 4-t-butylcyclohexyltrimethyl-

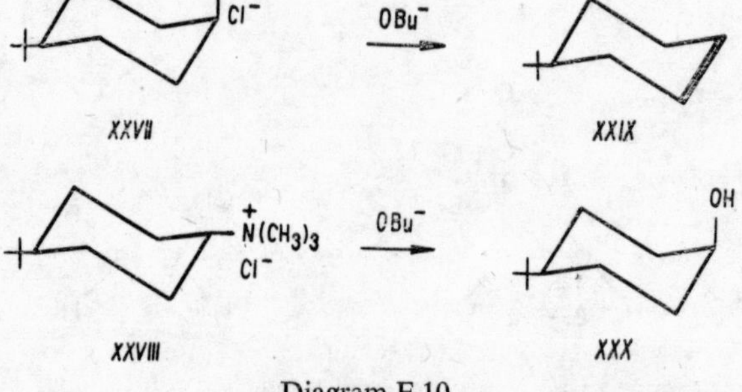

Diagram E 10

ammonium chloride with potassium t-butoxide, olefin *XXIX* is formed only from the *cis*-isomer, whereas the *trans*-isomer, having the trimethylammonium group attached in the equatorial position, gives the product of substitution *XXX* (Diagram E 10). The increase of nonbonded interactions accompanying the formation of the transition state also plays an important role. As the increase of non-bonded interactions in the transition state usually occurs during *cis*-olefin formation to a greater extent than during *trans*-olefin formation, *cis*-olefins are formed more slowly than the isomeric *trans*-olefins. For instance of the two

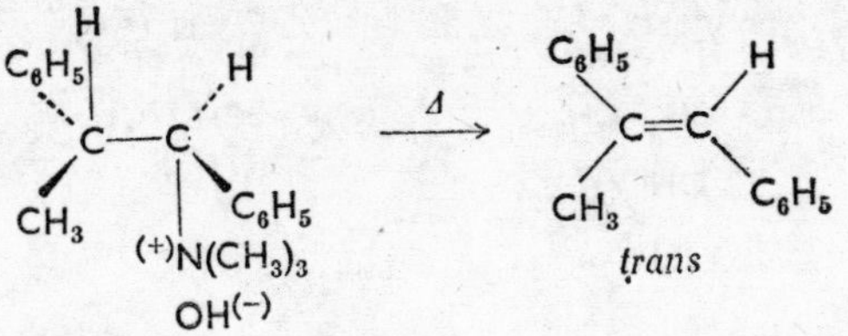

XXXI

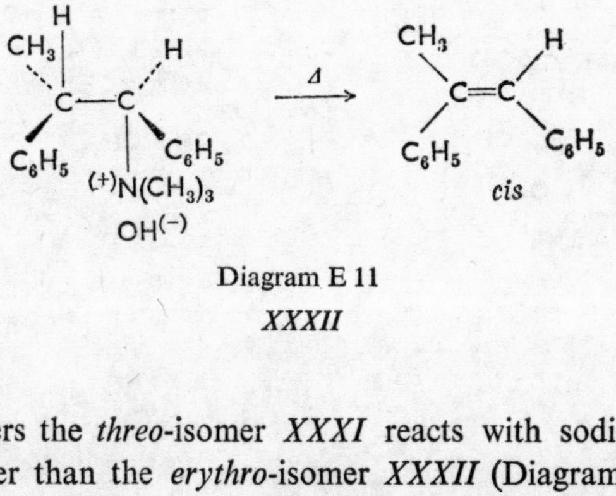

trans

cis

Diagram E 11

XXXII

diastereoisomers the *threo*-isomer *XXXI* reacts with sodium ethoxide 50 times faster than the *erythro*-isomer *XXXII* (Diagram E 11). The stereoelectronic condition of coplanarity applies also to the cleavage of quaternary hydroxides derived from *trans*-decahydroquinoline (*XXXIII*) and *cis*-octahydroindole (*XXXIV*), where the sterically favoured hydrogens are at $C_{(3)}$ in the first case and $C_{(7)}$ in the second case. On the other hand, ring contraction takes place (Diagram E 12) with the derivative of *trans*-octahydroindole (*XXXV*), where none of the hydrogens satisfy E_2 elimination conditions.

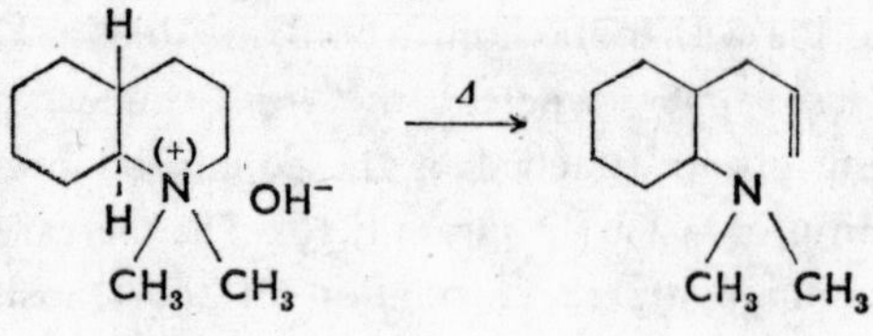

XXXIII

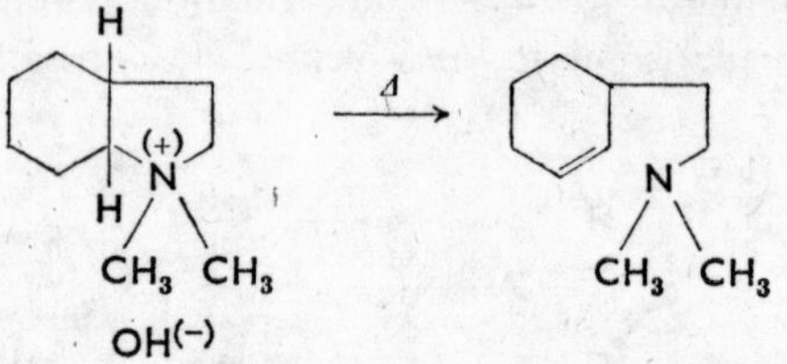

XXXIV

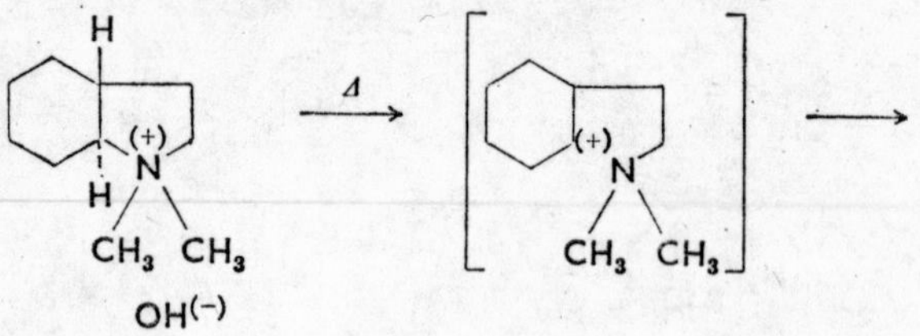

XXXV

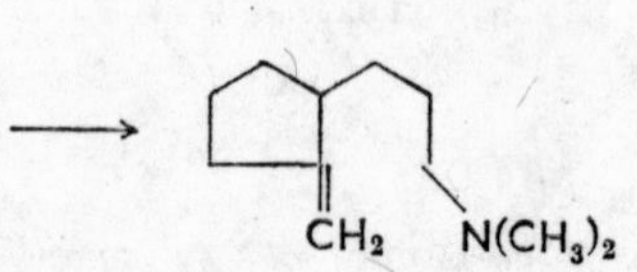

Diagram E 12

Cleavage of quaternary ammonium hydroxides of cyclic compounds up to a seven-membered ring yields *cis*-olefins exclusively, because *trans*-compounds are not capable of existence. In the case of cyclo-octyltrimethylammonium hydroxide, the conformations leading to *cis*-octene and *trans*-octene differ only slightly as regards energy, and therefore a mixture of both olefins is formed (see p. 16). Cyclononyltrimethylammonium hydroxide gives predominantly *trans*-cyclononene; in the ten-membered series only *trans*-cyclodecene is formed. Similarily, the cleavage of *cis*-cyclodecenyl-3-trimethylammonium hydroxide produces the more strained *cis-trans*-1,3-cyclodecadiene.

The cleavage of 2-hydroxytrimethylammonium bases of type *XXXVI* is not without interest. This reaction occurs also by an S_N2 mechanism and requires an *anti*-periplanar arrangement of the reacting C—OH and C—$N^{(+)}(CH_3)_3$ substituents in the transition state of the reaction. If these conditions are fulfilled, an epoxide is formed as the reaction product (Diagram E 13).

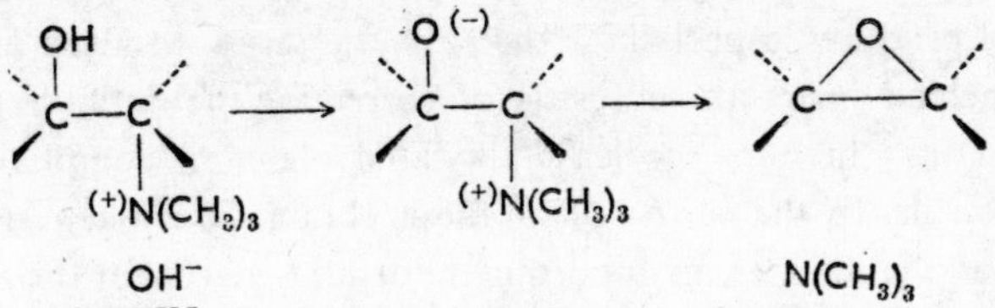

OH
C—C
$^{(+)}N(CH_3)_3$
OH^-
XXXVI

$O^{(-)}$
C—C
$^{(+)}N(CH_3)_3$

O
C—C
$N(CH_3)_3$

Diagram E 13

syn-Elimination

Of the *syn*-elimination reactions, the pyrolytic elimination of esters and amine *N*-oxides has recently gained importance. From a practical viewpoint, the reaction represents a means of introducing a double bond into an organic molecule under almost neutral conditions, thus limiting the possibility of a rearrangement in the course of the reaction (5).

Pyrolytic Cleavage of Esters

Esters of alcohols, if heated to 300 – 550 °C, decompose into an olefin and an acid. The reaction may be carried out in the liquid or preferably in the vapour phase. The simplest proof of a *syn*-elimination mechanism is the absence of elimination from compounds containing no *cis*-β-hydrogen with respect to the ester group. The study of steroid substances provided further evidence, on the basis of which a cyclic unimolecular mechanism (see Diagram E 14) has been proposed for this type of elimina-

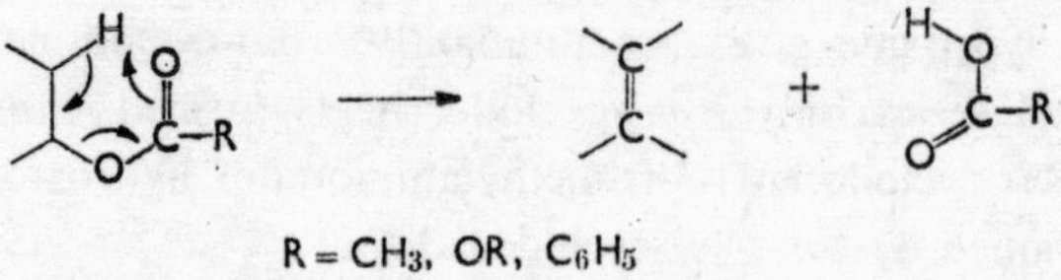

R = CH_3, OR, C_6H_5

Diagram E 14

tion. Benzoates, carbonates or sulphites are frequently used for the pyrolytic cleavage; the most important reaction, however, is the thermal decomposition of acetates or xanthates, known as the **Tschugaeff reaction** (6). The thermal decomposition of acetates of primary alcohols leads to single olefins. The decomposition of esters of secondary alcohols yields mixtures of products, especially in the aliphatic series. Modern analytical methods helped prove the necessity of correcting the initial hypothesis, that elimination favours the least alkylated olefin in compliance with Hofmann's rule. In the course of all these elimination reactions, especially in the aliphatic series, an important influence is exerted by the statistical factor, *i.e.* by the number of hydrogen atoms capable of *syn*-elimination attached to the individual β-carbons, as well as by steric effects, due to nonbonded interactions in the transition state. Furthermore, because

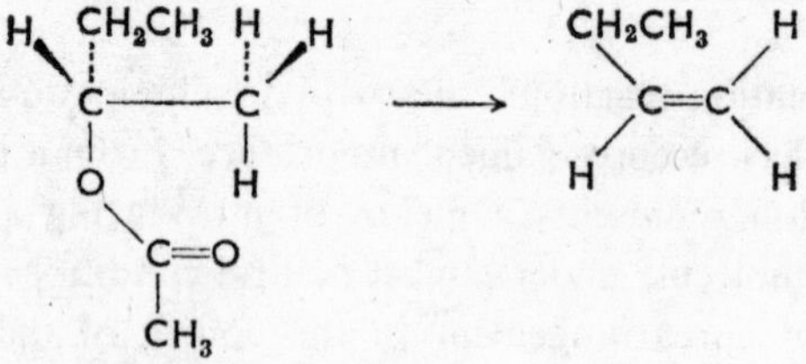

XXXVII

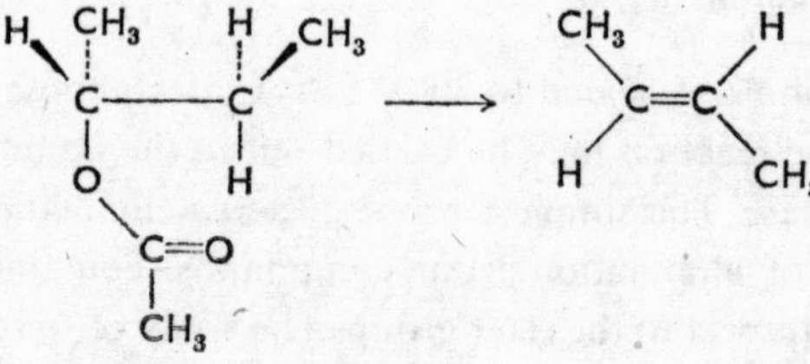

XXXVIII

Diagram E 15

the reactions take place at high temperatures, the thermodynamic stability of the olefins formed also asserts itself. The thermal decomposition of 2-butyl acetate, for instance, yields a mixture of 57 % 1-butene and 43 % 2-butene. Conformational analysis shows that during the elimination of hydrogen from the methyl group, all three hydrogen atoms may occupy the position required by the transition state of elimination *XXXVII*.

On the other hand, during the elimination of hydrogen from the methylene group only two hydrogen atoms come into consideration (Formula *XXXVIII*).

According to purely statistical evaluation, the reaction will lead to the formation of a mixture of 1-butene and 2-butene in a ratio of 3 : 2, which agrees very well with experimental results. Moreover it was established that the more substituted olefins are thermodynamically more stable. This will without doubt influence the composition of pyrolysis reaction products. For instance at 500 °C, the equilibrium mixture is composed of 28 % 1-butene, 41 % *trans*-2-butene and 31 % *cis*-2-butene.

In the case of more complex molecules, for instance 2,2-dimethyl-4-hexyl acetate, thermal decomposition ought to yield an equimolar mixture of 2,2-dimethyl-3-hexene and 2,2-dimethyl-4-hexene according to statistical evaluation. In fact the reaction mixture comprises more 3-alkene than 2-alkene and more *trans*-isomer than *cis*-isomer. In cases where the differences in energies between transition states are large, those with the fewest steric interactions are favoured. Elegant testimony to this fact is furnished by the thermal decomposition of the acetates prepared from *threo*- and *erythro*-isomers of 1,2-diphenylethanol-[2-D]. Almost all the deuterium from the *erythro*-isomer *XXXIX* was encountered in the stilbene obtained by elimination; conversely, the *threo*-isomer *XL* yielded only an insignificant proportion.

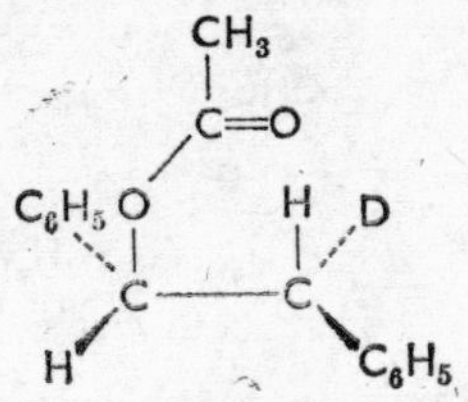

XXXIX

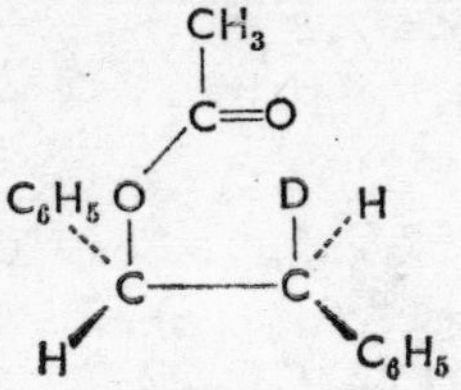

XL

In the case of alicyclic esters, olefins with an endocyclic double bond predominate in the reaction mixture. For example, the thermal decomposition of 1-methylcyclohexyl acetate (*XLI*) yields 75 % of 1-methylcyclohexene (*XLII*) and only 25 % of methylenecyclohexane (*XLIII*) – see Diagram E 16.

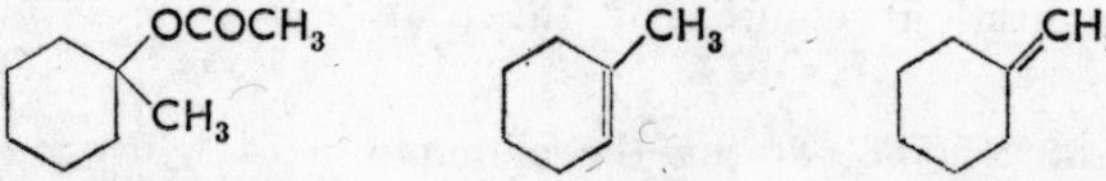

$OCOCH_3$ CH_3 CH_3 CH_2

XLI *XLII* *XLIII*

Diagram E 16

As the decomposition of esters takes place in the absence of acidic or basic agents it is especially suited for the preparation of very reactive dienes. Thus it was possible to prepare 4,5-dimethylenecyclohexene (*XLIV*) without causing isomerization to xylene (Diagram E 17).

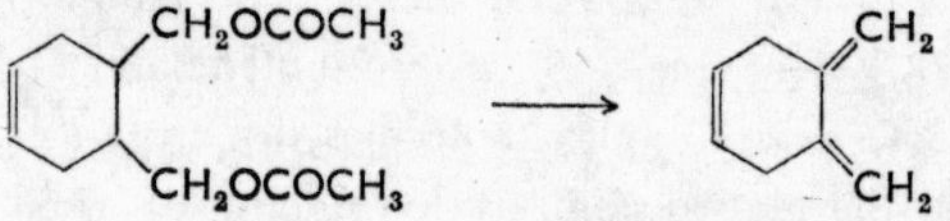

CH_2OCOCH_3 CH_2OCOCH_3 → CH_2 CH_2

XLIV

Diagram E 17

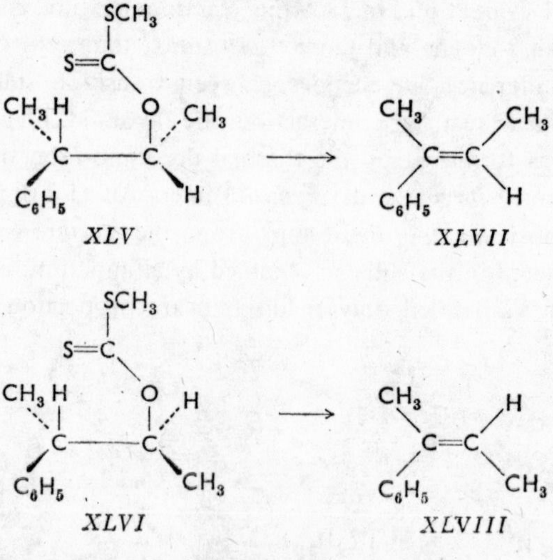

SCH_3 $S{=}C$ CH_3 H O CH_3 C—C C_6H_5 H → CH_3 CH_3 C=C C_6H_5 H

XLV *XLVII*

SCH_3 $S{=}C$ CH_3 H O H C—C C_6H_5 CH_3 → CH_3 H C=C C_6H_5 CH_3

XLVI *XLVIII*

Diagram E 18

The decomposition of xanthates resembles, as regards its mechanism, the thermal decomposition of acetates (7). For example, the thermal decomposition of the xanthates of the *threo*- (Formula *XLV*) and *erythro*- (Formula *XLVI*) isomers of 3-phenyl-2-butanol yields *cis*-2-phenyl-2-butene (*XLVII*) in the first case and *trans*-2-phenyl-2-butene (*XLVIII*) in the second case (see Diagram E 18). The stereospecificity of cleavage of xanthates is less marked than that of acetates. Up to 20 % of *anti*-elimination products are formed by the decomposition of isomeric decyl xanthates. The presence of an acidic hydrogen particularly influences the course of the elimination.

H, $SO_2C_6H_4CH_3$-*p*, H, $OCSSCH_3$ → $SO_2C_6H_4CH_3$-*p*

XLIX *L*

Pyrolytic Decomposition of Amine N-Oxides

The pyrolytic decomposition of amine *N*-oxides leading to an olefin and a substituted hydroxylamine (**Cope's reaction**) is also a *β*-hydrogen *syn*-elimination occurring, however, by the action of a base, as here we have an attack by an oxygen atom attached by means of a semi-polar bond to a nitrogen. *cis*-Dimethyl(2-phenylcyclohexyl)amine *N*-oxide, yields 3-phenylcyclohexene by decomposition. In the aliphatic series, *cis*-olefins are predominantly formed from *threo*-isomers and *trans*-olefins from *erythro*-isomers. The thermal decomposition of dimethyl(1-methylcyclohexyl)amine *N*-oxide yields up to 97 % of methylenecyclohexane whereas the decomposition of esters yields only 25 %. By comparing the thermal decomposition of esters and amine *N*-oxides we find that the decomposition of amine *N*-oxides is more stereospecific. Of preparative importance is the thermal decomposition of amine *N*-oxides containing medium rings. Cyclo-octyldimethylamine *N*-oxide yields only *cis*-cyclo-octene by thermal decomposition; the analogous reaction of nine- and ten-membered rings leads to *trans*-cyclo-olefins.

Other Examples of *syn*-Eliminations

A special case of *syn*-elimination is the reaction of ylides. Preparatively the most important example is the decomposition of ylides formed by the action of strong bases such as methyl-lithium or phenyl-lithium on quaternary ammonium salts (**Wittig reaction** (8)). In the case of heavily

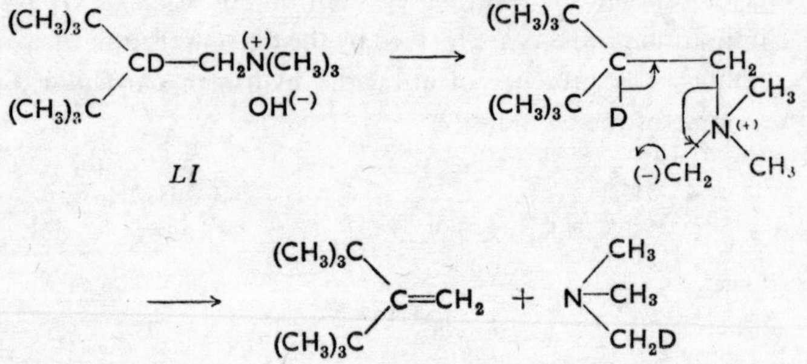

$$[(CH_3)_3C]_2CD-CH_2\overset{(+)}{N}(CH_3)_3\ OH^{(-)} \longrightarrow [(CH_3)_3C]_2C(D)-CH_2-\overset{(+)}{N}(CH_3)_2\overset{(-)}{C}H_2$$

LI

$$\longrightarrow [(CH_3)_3C]_2C{=}CH_2 + N(CH_3)_2(CH_2D)$$

Diagram E 19

sterically hindered bases, *e.g. LI,* the decomposition of ylides is preferred to the usual β-elimination (Diagram E19). The decomposition of 1,2-diphenylethyl acetate by the action of potassium amide is also a *syn*-elimination. Finally, alkyl halides when heated to a high temperature eliminate hydrogen halide and yield olefins. For example, the thermal decomposition of menthyl chloride (*LII*) yields 75 % of 3-menthene by means of hydrogen chloride *syn*-elimination.

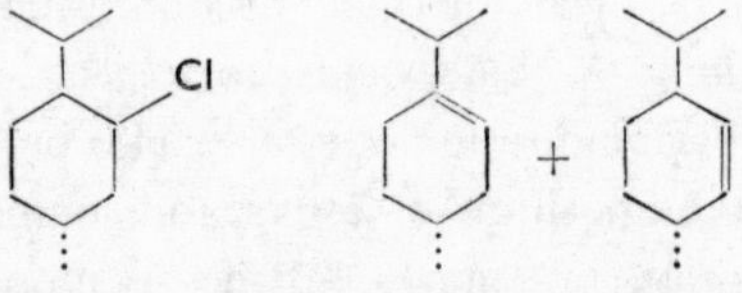

Cl

LII 3-(75 %) + 2-(25 %)

Diagram E 20

Literature

1. Brown H. C. *et al.*: J. Am. Chem. Soc. *78*, 2190, 2193, 2197, 2203 (1956).
2. Cristol S. J. *et al.*: J. Am. Chem. Soc. *74*, 2193 (1952); *79*, 3438, 3442 (1957).

3. Young W. G., Dillon R. T., Lucas H. J.: J. Am. Chem. Soc. *51*, 2528 (1929).
4. Cope A. C., Turmbull E. R.: *Organic Reactions,* Vol. XI, p. 317. Wiley, New York, 1961.
5. De Puy J., King R. W.: Chem. Revs *431* (1961).
6. Nace H. R.: *Organic Reactions.* Vol. XII, p. 57. Wiley, New York, 1962.
7. Wittig G.: Angew. Chem. *63*, 15 (1951).

F. THE STEREOCHEMICAL COURSE OF ADDITION REACTIONS

General Comments Concerning Mechanism

Addition reactions to unsaturated compounds may occur by either ionic or radical mechanisms. Isolated and unactivated double bonds between carbon atoms react easily with electrophilic reagents; the addition of nucleophilic reagents takes place only if the double bond is activated by groups strongly attracting electrons ($-CHO$, $-COR$, $-COOR$, $-NO_2$, $-CN$, *etc.*):

$$\mathrm{R{-}CH{=}CH_2 + X\cdot \,|\, \cdot Y \rightarrow \begin{array}{c}\mathrm{R{-}CH{-}CH_2}\\ \quad\;\; | \quad\;\; | \\ \quad\;\; \mathrm{Y} \quad\;\, \mathrm{X}\end{array}} \qquad \text{(radical addition)}$$

$$\mathrm{R{-}CH{=}CH_2 + X \,|\, {:}Y \rightarrow \underset{(+)}{\mathrm{R{-}CH}}{-}CH_2X} \qquad \text{(ionic electrophilic addition)}$$

$$\mathrm{\underset{(+)}{R{-}CH}{-}CH_2X + Y{:} \rightarrow \begin{array}{c}\mathrm{R{-}CH{-}CH_2}\\ \quad\;\; | \quad\;\; | \\ \quad\;\; \mathrm{Y} \quad\;\, \mathrm{X}\end{array}}$$

All these addition reactions, although they appear very simple at first sight, have a very complex kinetic course. The addition of the electrophilic particle determines the rate of the whole process and it depends on the nature of the electrophile, on the structure of the molecule at which reaction occurs and, finally, on the reaction medium, how the reaction will proceed. Ionic additions to non-symmetrically substituted double bonds are governed by the **Markovnikov rule**, radical additions by the **Kharasch rule**. From the stereochemical aspect, addition reactions to double bonds between carbon atoms can be either *syn*-additions, when both particles are added from the same side of the original double bond (see Diagram F 1), or *anti*-additions, when each of them is added from opposite sides (see Diagram F 2).

A reaction may be subject to **kinetic control**, in which case the ratio of products is determined by their relative rates of formation, or alternatively to **thermodynamic control**, when the proportions of the isomers are determined by the thermodynamic stabilities of the final reaction products.

Diagram F 1

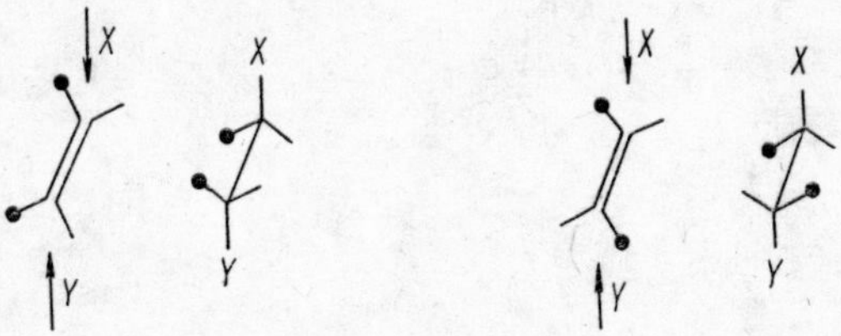

Diagram F 2

Additions to multiple bonds between different atoms (C = O, C = NH, *etc.*) are mostly nucleophilic. The addition of a nucleophile occurs at the site of lower electron density and determines the overall rate of the reaction.

Ionic Addition Reactions to Double and Triple Bonds between Carbon Atoms

anti-Additions of Halogens, Hypochlorous Acid and Hypobromous Acid

The addition of a halogen to a double bond is a stereoselective reaction. Since the addition is electrophilic and the addition of the bromonium ion determines the overall rate of the reaction, its stereoselective course cannot be explained by the formation of the classical carbonium ion, which would lead to a mixture of products because of the possible rotation around the single C—C bond being formed. We therefore assume that

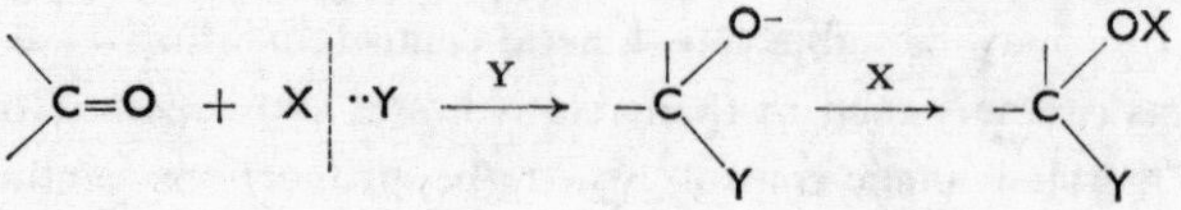

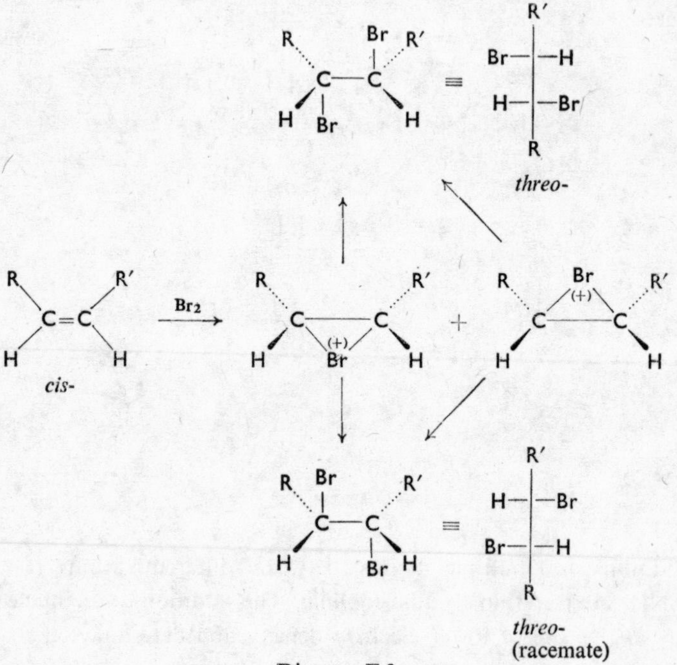

$$>C{=}O + X|{:}Y \xrightarrow{Y} -C(O^-)(Y) \xrightarrow{X} -C(OX)(Y)$$

R, Br, R′, C—C, H, Br, H ≡ R′, Br—H, H—Br, R

threo-

R, R′, C=C, H, H

cis-

$\xrightarrow{Br_2}$

R, R′, C—C, H, (+) Br, H + R, Br (+), R′, C—C, H, H

R, Br, R′, C—C, H, Br, H ≡ R′, H—Br, Br—H, R

threo-
(racemate)

Diagram F 3

one of the three free electron pairs of the halogen takes part in the formation of a bond with the developing carbonium ion, thus forming a chloronium, bromonium or iodonium cation with a three-membered ring and preventing rotation around the C—C bond. The geometrical arrangement of a halonium ion, where all four substituents of the original double bond are contained in one plane perpendicular to the plane of the three-membered ring formed, facilitates the subsequent reaction with the halogen anion which is connected with an inversion of the configuration at the attacked carbon atom. Thus *trans*-olefins yield 1,2-dihalogen derivatives of *erythro*-configuration and *cis*-olefins yield compounds of

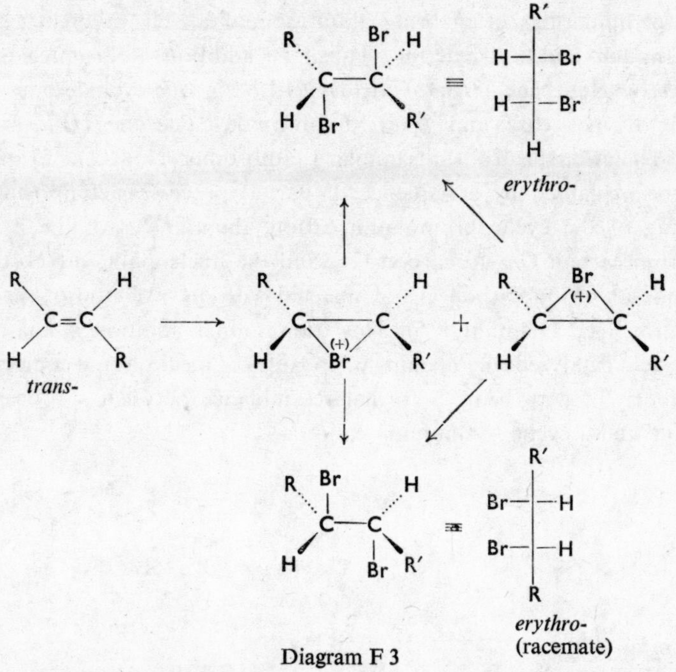

Diagram F 3

threo-configuration. If the olefin contains a plane of symmetry then two enantiomeric bromonium ions are formed (see Diagram F 3) in equal amount and the product is a racemate. Had the double bond been symmetrically substituted, *meso*-products would result in the former case and racemates in the latter (see Diagram F 3).

Hypochlorous and hypobromous acids may add to double bonds in a similar manner. For example, cyclohexene (*I*) yields the racemic *trans*-bromohydrin *II*, which may easily be converted into an epoxide (see Diagram F 4). The extraordinary steric conditions of cycloalkenes

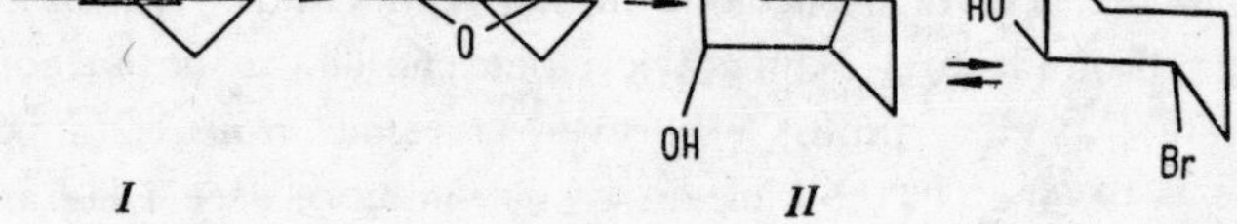

Diagram F 4

with medium rings often bring about a qualitatively different type of transannular addition reaction. Thus, the addition of bromine to *cis*- or *trans*-cyclodecene (*IIIa, b*) yields *cis*-1,6-dibromocyclodecane (*IVa*) in the former case and *trans*-1,6-dibromocyclodecane (*IVb*) in the latter. The mechanism of transannular 1,6-dibromocyclodecane formation may be explained by a sequence of partial processes, comprising the opening of the cyclic bromonium cation, the transfer of the hydride ion from carbon $C_{(6)}$ to carbon $C_{(2)}$ and the nucleophilic attack of the bromide anion on carbon $C_{(6)}$ connected with an inversion of the configuration (see Diagram F 5). This transannular addition is analogous to the acid catalysed ring opening of epoxides of medium ring compounds (see p. 197), there being a formal resemblance between a protonated epoxide and a cyclic bromonium cation.

$Br^{(-)}$ H $Br^{(+)}$ Br Br

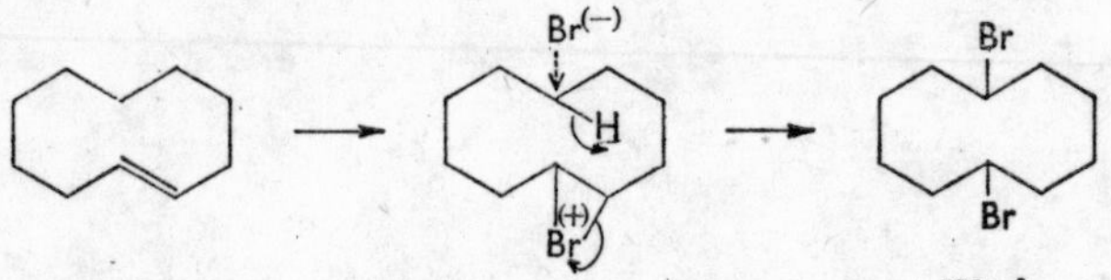

IIIa,b *IVa,b*

Diagram F 5

anti-**Additions of Hydrogen Halides and Water**

Compounds containing isolated double bonds add hydrogen halides, and indirectly also water with the help of sulphuric acid, by an *anti*-mechanism according to the Markovnikov rule. The stereospecific course of these additions is explained by the formation of π-complexes from the proton and the olefin followed by addition of the nucleophilic particle. In the case of hydration reactions, the strongly polar π-complex orientates the surrounding water molecules as a result of electrostatic forces, thus preventing possible rotation around the arising C—C single bond. The steric course of the ionic addition of hydrogen halides was studied mainly in the field of cyclic unsaturated compounds. The addition of hydrogen bromide to 1,2-dimethylcyclohexene (*V*) in hexane or acetic acid at 0 °C yields almost exclusively racemic *trans*-1,2-dimethyl-1-bromocyclohexane (*VI*) by means of *anti*-addition (see Diagram F 6). However, the configuration of the product of addition in acetic acid is

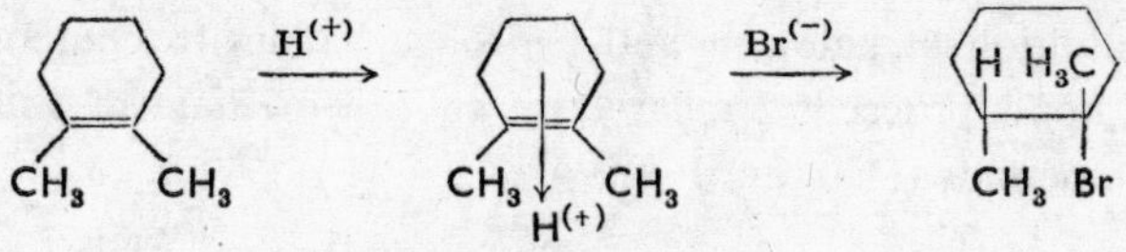

Diagram F 6

significantly influenced by the reaction temperature and the reaction time. Already after six minutes, 12 % of *cis*-1,2-dimethyl-1-bromocyclohexane is present in the reaction mixture, approaching thermodynamic equilibrium (85 % *trans*- and 15 % *cis*-bromo derivative). The addition of hydrogen halides to α, β-unsaturated acids is also an *anti*-addition. The addition of hydrogen bromide to the isomeric pair of acids, tiglic acid (*cis*, *VII*) and angelic acid (*trans*, *VIII*), leads to the β-bromo acid of *erythro*-configuration in the former case and to the *threo*-isomer in the latter. Such marked sterospecificity may be explained by assuming the formation of a π-complex and its opening from the opposite side to that where the proton is located (see Diagram F 7). The configuration was confirmed by the fact that in the first case *trans*-2-butene (*IX*) is formed from the acid by the action of base with the simultaneous elimination of the halogen and carbon dioxide, whereas *cis*-2-butene (*X*) is the product in the second case.

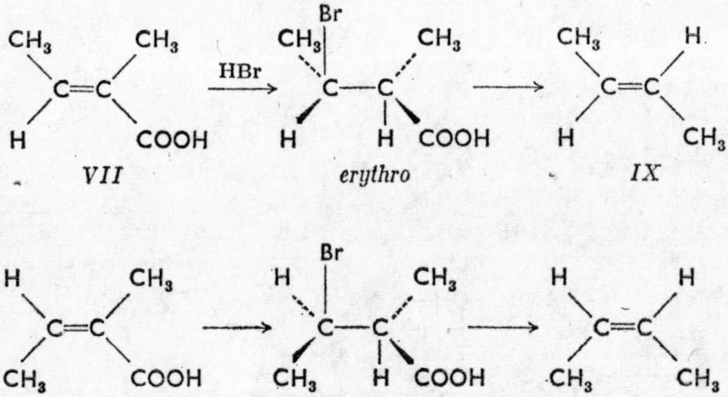

Diagram F 7

The addition of halogens and hydrogen halides to compounds with a triple bond proceeds similarly by an *anti*-mechanism and leads to *trans*-compounds (Diagram F 8).

HOOC—C≡C—COOH ⟶ HOOC(Br)C=C(Br)COOH

Diagram F 8

syn-Additions

In addition to the examples of *anti*-additions discussed above, there is a number of reactions occurring by *syn*-mechanism. We will not be guilty of a grave error if we state that all *syn*-additions presume the formation of a cyclic structure, be it an isolatable intermediate or only a transition state, which determines the configuration of the reaction products.

One of the reactions of preparative importance is the 1,2-addition of two hydroxyl groups to a double bond, usually called hydroxylation of a double bond. Hydroxylation by means of a dilute alkaline solution

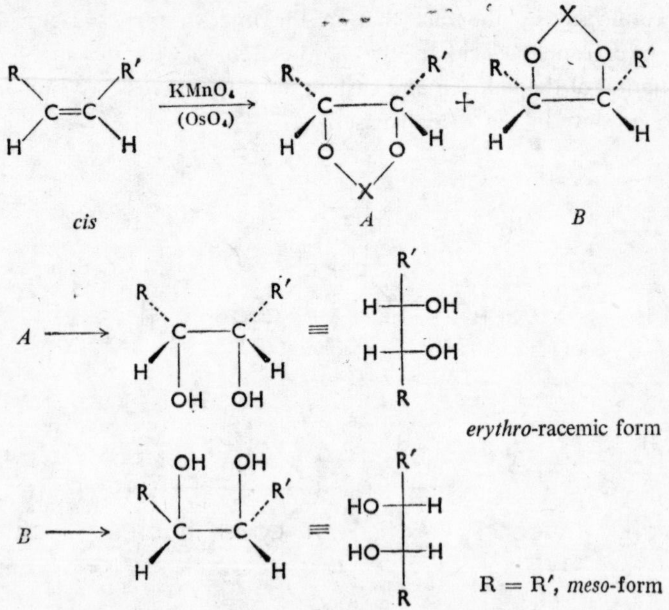

cis ⟶ ($KMnO_4$ (OsO_4)) A + B

A ⟶ ≡ erythro-racemic form

B ⟶ ≡ R = R′, *meso*-form

Diagram F 9

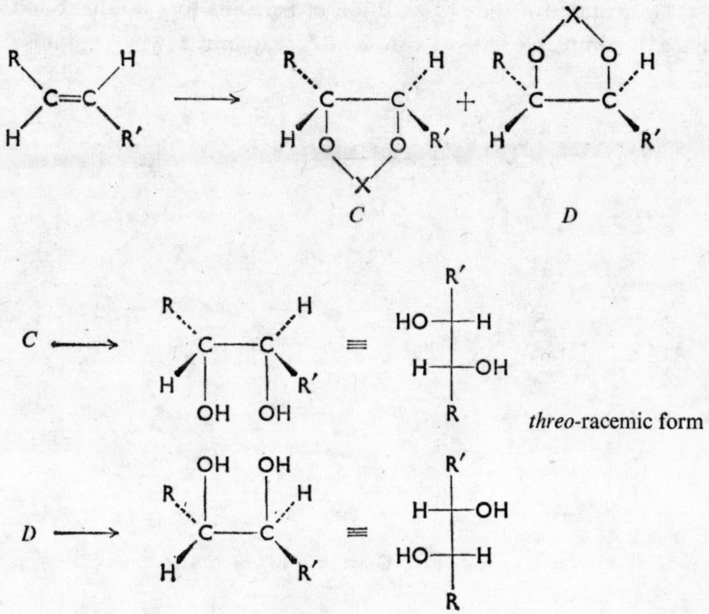

threo-racemic form

Diagram F 9

of potassium permanganate (1) or with the help of osmium tetroxide occurs *via* cyclic esters, which yield α-glycols upon hydrolytic decomposition. By means of this procedure we obtain α-glycols of *erythro*-configuration from *cis*-olefins, whereas glycols of *threo*-configuration are formed from *trans*-olefins (see Diagram F 9). Maleic acid thus yields mesotartaric acid; fumaric acid gives racemic tartaric acid. Similarly, *cis*-crotonic acid gives the $(\pm)$-*erythro*-form of 2,3-dihydroxybutyric acid, and *trans*-crotonic acid yields the $(\pm)$-*threo*-form. If there is an asymmetric carbon attached to the double bond in its immediate surroundings, asymmetric addition takes place, favouring the side sterically more accessible (see p. 221).

Another important reaction is the formation of epoxides by the *syn*-addition of organic peracids to a double bond (see Diagram F 10). The fact that the reactivity of peracids increases with their increasing acidity ($CF_3COOOH > HCOOOH > CH_3COOOH$) testifies in favour of the hypothesis that the reaction is an electrophilic addition. The transition state for the *syn*-addition of peracids to a double bond resembles a π-complex. Thus it can also be explained why α, β-unsaturated

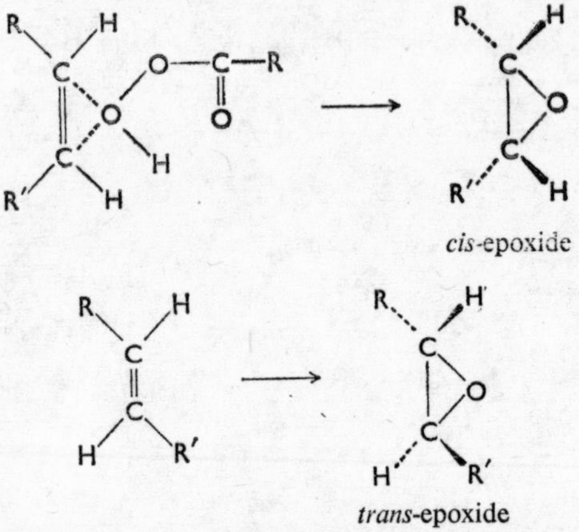

cis-epoxide

trans-epoxide

Diagram F 10

ketones, acids and their derivatives are very reluctant to react with peracids. For example, 2-methylbutene reacts at a 7,500 times greater rate with peracetic acid than does cinnamic acid.

The epoxides are configurationally identical with the products obtained by the addition of hypochlorous acid or hypobromous acid to olefins (*anti*-addition) and by subsequent dehydrohalogenation (*anti*-elimination) of the halogenohydrins formed. The alkaline as well as acidic decomposition of epoxides yields α-glycols. Since these reactions again occur stereospecifically (see p. 191), we obtain the *threo*-form from the *cis*-olefins and the *erythro*-form of the α-glycols from the *trans*-olefins. From symmetrical olefins, a racemate is formed in the first instance and a *meso*-form in the second. By means of this procedure, we are able to prepare α-glycols which differ in configuration from those prepared by *cis*-hydroxylation with the help of potassium permanganate or osmium tetroxide. Hydrogen peroxide, in the presence of various oxides, for example vanadium pentoxide, chromium trioxide or, preferably, selenium dioxide, tungsten trioxide and molybdenum trioxide, produces directly the products of the *trans*-addition of two hydroxyl groups to a double bond. In some cases the opening of the epoxide ring occurs in an anomalous manner. For example, norbornene (*XI*) treated with performic acid yields the 2,7-glycol *XII* by means of a reaction proceeding *via* a non-classical ion (see Diagram F 11). The epoxides of cycloalkenes with medium rings react almost exclusively transannularly in the course of acid-catalysed ring opening, in contrast to normal and large rings.

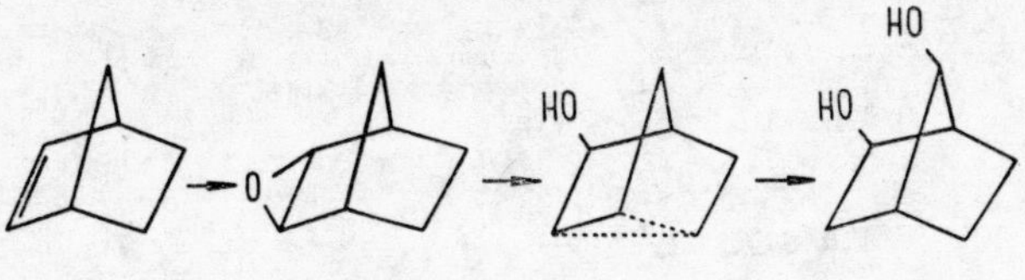

HO HO HO

XI *XII*

Diagram F 11

In this case, the nucleophilic reagent does not react with the carbons participating in the formation of the epoxide but with the opposite part of the ring, with a shift of the hydride ion. 1,2-Epoxycyclo-octane gives 1,4-dihydroxycyclo-octane (see Diagram F 12); 1,2-epocycyclononane yields 1,5-dihydroxycyclononane. 1,6-Dihydroxycyclodecane is formed in the ten-membered ring series.

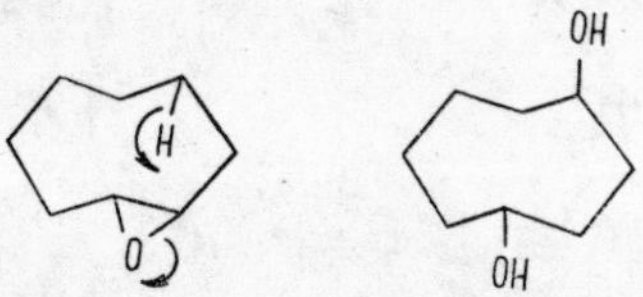

OH H O OH

Diagram F 12

Hydroxylation reactions also include the additions of lead tetra-acetate or silver acetate and iodine (the **Prevost reaction**) to the double bond. In the latter case, the reaction is started by a *trans*-addition but leads to the same cyclic intermediate as in the former case. By opening the ring of the intermediate in the presence of water we arrive at the glycols accessible also by direct *cis*-addition (see Diagram F 13). Conversely, stronger nucleophilic reagents yield *trans*-addition products (see p. 155).

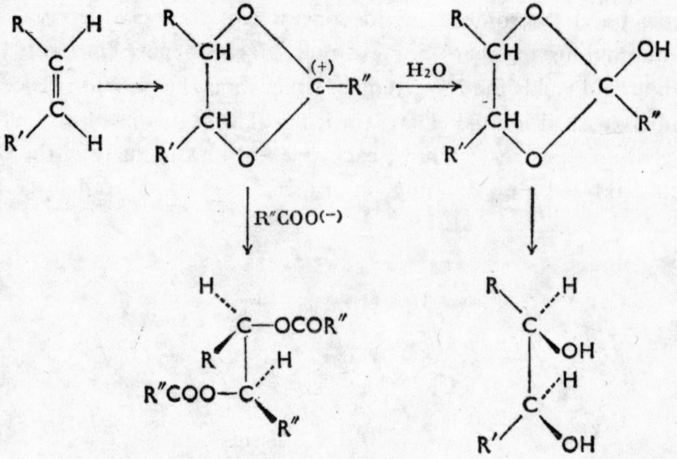

Diagram F 13

A typical classical example of *cis*-addition is the **Diels-Alder reaction** or **diene synthesis**, occurring by a *syn*-mechanism with regard both to the diene and to the dienophile. The relative positions of the substituents of the dienophilic component are maintained in the product as well. The *cis*-compound gives rise to a *cis*-compound, and the *trans*-compound yields a *trans*-compound (2), as is shown in Diagram F 14. In the course

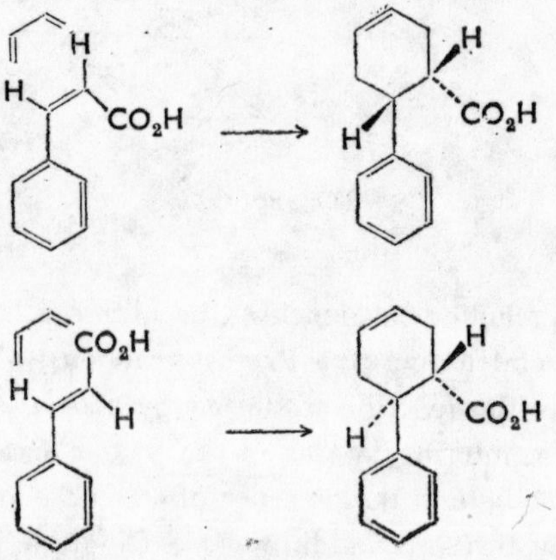

Diagram F 14

of diene synthesis with cyclic dienes, two isomeric products of *endo*- and *exo*-configurations, respectively, may be formed. The former usually predominate in the reaction mixture (Diagram F 15).

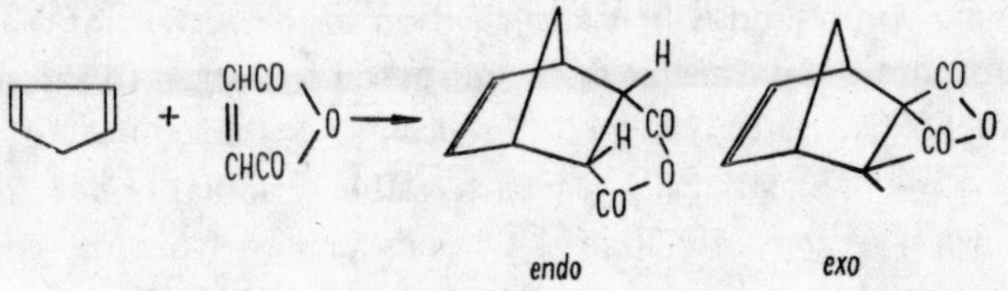

CHCO
CHCO
O
H
CO
H
CO
O
CO
CO
CO
O

endo *exo*

Diagram F 15

Further examples of *syn*-additions are the addition of diborane to a double bond, as has been proved on steroid compounds, and the reaction of olefins with carbenes obtained either by the photochemical reaction of aliphatic diazo compounds or by the action of strong bases on trihalogenomethanes. For example, the addition of dibromocarbene CBr_2 to *cis*- or *trans*-2-butene yields the *cis*-derivative *XIII* of 1,2-dimethylcyclopropane in the first case and the *trans*-derivative *XIV* in the second case.

CH_3 CH_3
Br
H H
Br

XIII

CH_3 H
Br
H CH_3
Br

XIV

Radical Addition Reactions to Double and Triple Bonds between Carbon Atoms

Radical additions are usually catalysed by the presence of metal catalysts, peroxides, or ultra-violet light, or by providing suitable reactive radicals. Under these conditions, the radical addition of the following substances takes place: hydrogen, bromine, hydrogen bromide, ethyl bromoacetate, bromotrichloromethane; sulphur-containing compounds hydrogen sulphide, mercaptans, thiophenols, thio acids; and the less usual reagents nitrogen dioxide, trichlorosilane, oxygen, sulphur dioxide, *etc*. (3).

Catalytic Hydrogenation

Isolated and unactivated double bonds may be reduced by means of catalytic hydrogenation. We assume that during hydrogenation the unsaturated compound is first adsorbed on to the surface of the catalyst and then follows the transfer of the adsorbed hydrogen to the substrate. Accordingly, hydrogen *syn*-addition occurs in the course of catalytic hydrogenation. The sterically more accessible terminal bonds are hydrogenated with greater ease than the bonds farther from the end of the chain or multiple bonds accommodated in a ring. With increasing number of groups attached to the double bond hydrogenation is slowed down; tetrasubstituted double bonds are hydrogenated least readily or not at all. The partial hydrogenation of the triple bond in cycloalkynes is a suitable method for the preparation of *cis*-cycloalkenes with medium and large rings.

Radical Addition of Compounds Containing Bromine

Hydrogen bromide and bromine may, by means of reversible addition to the double bond, bring about partial or total isomerization of alkenes before the addition proper, thus impairing the investigation and evaluation of the course of addition. Such a complication may be removed in the case of cyclic compounds, and therefore the steric course of radical reactions has mostly been studied on them. The addition of hydrogen bromide to 1-bromocyclohexene (*XV*) or to 1-methylcyclohexene (*XVI*) gives, as almost the only product, the thermodynamically less stable *cis*-1,2,-dibromocyclohexane (*XVII*) or *cis*-1-methyl-2-bromocyclohexane (*XVIII*), respectively. Similarly, the *cis*: *trans* ratio of the 1,2-dibromo derivatives obtained by the radical addition of hydrogen bromide to 1-bromocyclobutene, 1-bromocyclopentene and 1-bromocycloheptene (79 : 21, 94 : 6, 91 : 9) testifies to the stereoselective course of the addition. All the examples referred to are *anti*-additions, in the course of which the atomic bromine formed by the homolytic cleavage of hydrogen bromide reacts first of all with the olefin, giving rise to a radical of similar structure as the bromonium ion has in the course of ionic reactions. The *cis*-isomer is then formed by means of a subsequent reaction with hydrogen (Diagram F 16).

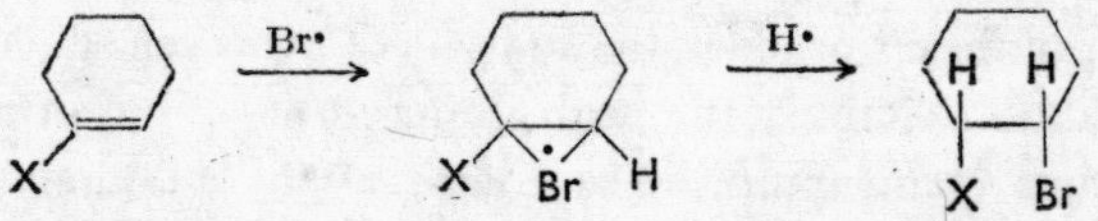

XV, X = Br
XVI, X = CH_3

XVII, X = Br
XVIII, X = CH_3

Diagram F 16

It is also possible that in the primarily formed radical with the axial bromine atom (Formula *XIX*) or the equatorial one (Formula *XX*), the access of hydrogen bromide from the opposite side to the bromine atom is sterically favoured in the latter case.

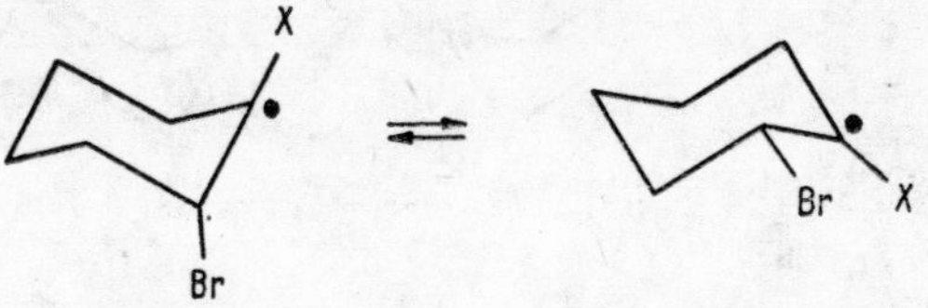

XIX *XX*

The radical addition of hydrogen bromide or deuterium bromide to *cis*- or *trans*-2-bromo-2-butene at −80 °C is also stereoselective. In the first case mainly *meso*-2,3-dibromobutane has been isolated; in the second one only the racemic *threo*-derivative has been encountered. However, a mixture of both isomers is formed at higher temperatures. The steric course of the addition of hydrogen bromide to a triple bond depends to a great extent upon the steric requirements and polarity of the alkyls attached to the atoms of the triple bond.

Radical Addition of Compound Containing Sulphur

The radical additions of hydrogen sulphide, thiophenols and thio acids to olefins are not as stereoselective as those described above. Significant selectivity may be observed only in the case of double bonds with a marked steric hindrance. The evaluation of these additions is complicated mainly by the fact that the ratio of the isomers obtained by *syn*- or *anti*-addition depends on the ratio of the starting materials. The addition of *p*-thiocresol to norbornene yields the *exo*-isomer only. On the other hand, the addition of thiophenol to 1-chlorocyclohexene gave 92.4 % of the *cis*-product, the addition of hydrogen sulphide 74.8 %, and the addition of thioacetic acid produced 66–73 % of the product with the same configuration. The course of the addition is evidently connected with the stability of the primarily formed radical. If we assume that the atom of sulphur in a 2-substituted 1-chlorocyclohexyl radical is localized in the axial position and that addition occurs before conformational equilibrium is established, we obtain *cis*-product *XXI* by means of *anti*-addition; if on the other hand conformational equilibrium is established, we may expect the formation of both products, *XXI* as well as *XXII* (see Diagram F 17).

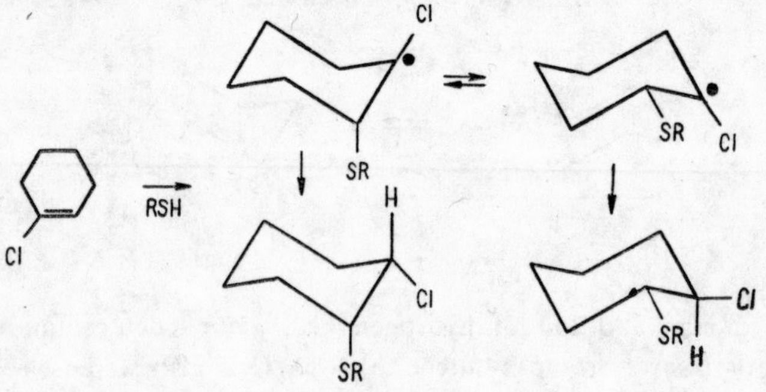

Cl
RSH
Cl
SR
SR Cl
H
Cl
SR
Cl
SR
H
XXI
XXII

Diagram F 17

The radical addition of nitrogen tetroxide to olefins yields the nitrite esters of 2-nitroalcohols. Cyclohexene thus gave 58 % of the nitrite of *trans*-2-nitrocyclohexanol in addition to 42 % of the *cis*-isomer, whereas 1-methylcyclohexene yielded the nitrite of *trans*-1-methyl-2-nitrocyclohexanol only. This significant difference is connected with the stability of the primarily formed radical, as in the preceding case.

The Steric Course of Addition Reactions to the Carbonyl Group

The carbonyl group in aliphatic compounds forms a component of a mobile chain which may occupy a number of conformations. Disregarding polar effects, the ease of addition of nucleophilic particles is here influenced by the steric requirements of the substituents located in the near vicinity of the carbonyl group. Such substituents may hinder or prevent completely (in the case of di-t-butylketone) the addition of nucleophilic particles. The ease of addition of nucleophilic reagents to the carbonyl group of cyclic ketones is influenced by the conformation of the latter. The increased reactivity of cyclopropanone and cyclobutanone is caused by the deformation of the bond angles. Nonbonded interactions influence the reactivity of cyclopentanone and cyclohexanone, which is smaller in the former case. As a result of reduced mobility and thus also of a smaller number of conformations, reactions at the carbonyl group of alicyclic compounds occur stereoselectively. Whilst catalytic hydrogenation of cyclohexanone derivatives, especially in an acidic medium, leads to alcohols with the hydroxyl group in axial position, reduction with the help of metals and complex hydrides (Na, $NaHg_x$, $LiAlH_4$, $NaBH_4$, *etc.*) gives predominantly hydroxy compounds with the hydroxyl in equatorial position (Diagram F 18). Analogous conclusions are also valid

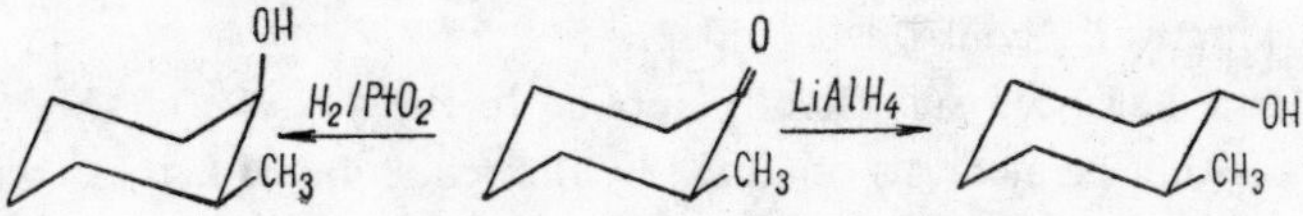

Diagram F 18

for the reduction of oximes of cyclic ketones. A new asymmetric carbon is formed in the course of all these additions to the carbonyl group, with the exception of symmetrical ketones as starting materials. If a chiral centre is present in the vicinity of the carbonyl group, so-called **asymmetric induction** affects the course of the addition of nucleophilic reagents. By this term we mean the influencing of the formation of a new chiral carbon atom by a chiral atom already present in the molecule (for further details see p. 212). In 1952, Cram and Elhafez suggested a very useful generalization (4) for non-catalytic processes of this type, the so-called **rule of steric control of asymmetric induction**. The original ketone reacts in the conformation in which the double bond of the carbonyl group is placed between the smallest group and the medium size one of the three groups attached to the neighbouring carbon. The nucleophilic group approaches the carbonyl group from the less hindered side, *i.e.* from the side where the smallest substituent of the chiral atom is located, so that the formation of one of the two possible diastereoiso-

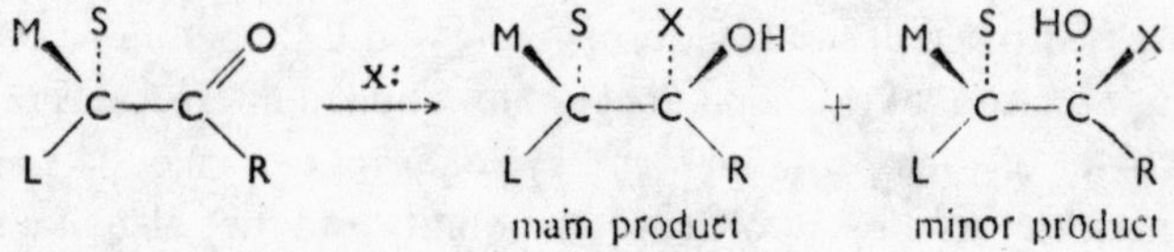

Diagram F 19

meric compounds predominates (Diagram F 19). (L designates the sterically large substituent, M the medium one and S the small one.)

The formulation of the rule already includes the reservation that the rule is valid only for processes in which the influence of a catalyst is not manifest. The validity of the rule is further limited to reactions with a kinetically controlled course.

The rule has been successfully applied to the analysis of the ratio of diastereoisomeric alcohols obtained by the reduction of ketones with the help of complex hydrides, alkali metals and sodium amalgam, as well as in the course of the addition of organometallic reagents. The stereospecificity of the reaction is lower in the case of organolithium reagents than with organomagnesium reagents, because the former have lower coordination and association capacities, so that their effective volume is lower and steric effects do not exert such an influence. In the case of cyclic ketones, the Cram rule may be used to a certain extent only, because the relative differences between the possible transition states are not determined by the interactions between α-substituents and the particle being added but by the overall conformational arrangement of the whole compound. The Cram rule has also been used for the analysis of the steric course of the Wallach and Leuckart reduction of aliphatic and cyclic ketones, where the C=N bond is reduced in the course of the reaction. 2-Substituted cyclohexanones (*XXIII*; R=CH_3, C_2H_5) yield predominantly *cis*-1-alkyl-2-aminocyclohexanes (*XXIV*, Diagram F 20).

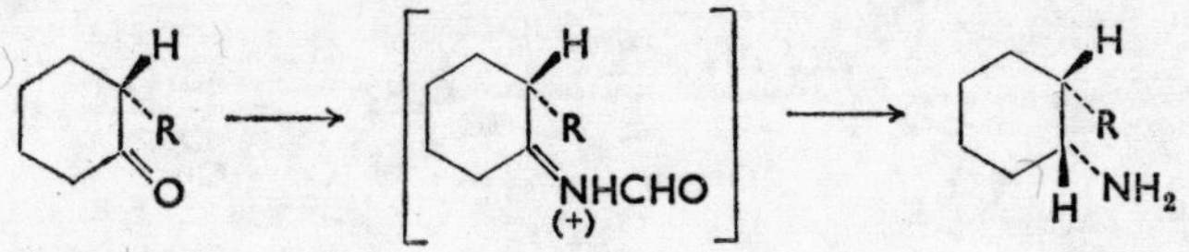

XXIII XXIV

Diagram F 20

The stated conclusions regarding reactions of carbonyl compounds are valid for compounds not accommodating any functional group in the molecule other than the carbonyl group. In the case of compounds carrying a group capable of interaction with the reagent (—OH, —NH_2) in the vicinity of the carbonyl (in positions α or β), we may consider an alternative reaction mechanism. For the cyclic transition state *XXV* we have to expect, because of its greater rigidity, a more unequivocal orientation of the attacking reagent and thus also a more stereospecific course than was true of the non-cyclic mechanism.

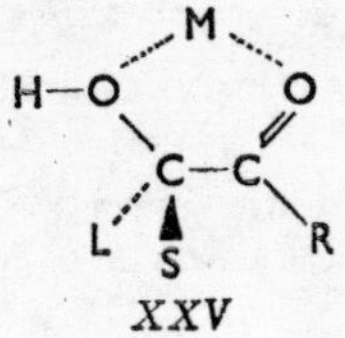

XXV

Reactions in the course of which the asymmetric atom influences the steric course of addition reactions at the carbonyl group and leads to the formation of optically active substances will be discussed in the following chapter pertaining to the course of asymmetric reactions (see p. 207). A further interesting reaction is the **Claisen reaction** between benzaldehyde and the enol form of phenylacetic acid (Diagram F 21). Of the two possible transition states, *XXVI* and *XXVII*, the former has a lower energy content as a result of smaller steric interactions, and therefore prevails. The reaction products contain a larger percentage of the *threo*-isomer (76 %, *XXVIII*) than of the *erythro*-isomer (24 %, *XXIX*) of 2,3-diphenyl-3-hydroxypropionic acid.

Diagram F 21

Literature

1. Wiberg K. B., Saegebarth K. A.: J. Am. Chem. Soc. *79*, 2822 (1957).
2. Alder K., Stein G.: Angew. Chem. *47*, 837 (1934); *50*, 510 (1937).
3. Bohm B. A., Abell P. I.: Chem. Revs. 599 (1962).
4. Cram D. J., Abd Elhafez: J. Am. Chem. Soc. *74*, 5828 (1952).

G. ASYMMETRIC REACTIONS LEADING TO OPTICALLY ACTIVE COMPOUNDS

The term **asymmetric reaction** in the broadest sense includes all reactions leading to the formation or destruction of a chiral centre which proceed to a different extent or at a different rate for the two possible configurations at this centre. The term "asymmetric reaction", defined in this manner, merges with the term "stereospecific reaction". In the narrower sense of the word, we understand the term as including only those reactions in the course of which an optically active compound is formed from a compound originally optically inactive. In the following chapter we will keep to the latter definition. Special attention has to be given to the efforts aimed at determining the configurational relationships between the starting materials and products. This is today the most frequent application of asymmetric reactions.

The majority of the reactions discussed in this chapter could have been included in the chapter referring to asymmetric induction. However, we consider the separate analysis of the reactions leading to optically active compounds preferable mainly from the teaching point of view.

Characteristic Features of Asymmetric Reactions and their Classification

The asymmetric course of a reaction is conditioned by the presence of at least one asymmetric component in the reaction system, as has already been expressed in principle by Pasteur. According to the nature of the asymmetric factor we may distinguish two types of reactions: a) reactions occurring under the influence of an asymmetric factor of physical nature, and b) reactions in which one of the reaction components already has a chiral molecular structure.

The first group comprises a small number of **photoreactions occurring under the influence of circularly polarized light**. In the course of such reactions, the selective formation of destruction of one of the possible configurations at a chosen centre of chirality takes place. W. Kuhn established with simple reactions that light of the wavelengths corresponding to the absorption bands in the ultra-violet spectrum of the starting material was most effective (1). The photochemical decomposition of *N,N*-dimethyl-α-azidopropionamide (*I*) may serve as an example. The decomposition occurs at a greater rate with one enantiomer than with

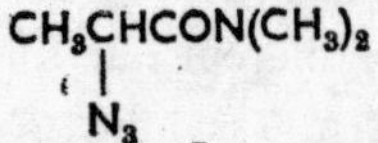

$$\underset{\displaystyle N_3}{CH_3\overset{|}{C}HCON(CH_3)_2} \qquad I$$

the other, so that the unchanged amide is optically active. An interesting example is the photodehydrogenation of the dihydropyridine derivative *II* leading to the asymmetric product *III*, the activity of the latter being caused by atropisomerism of the biphenyl type (Diagram G 1). In the course of the reaction, the elimination of the asymmetric carbon takes place to a different extent for the two possible configurations. The configuration of the atropisomer formed is related to the configuration of the more reactive enantiomer.

NO_2 H CH_3CO $COOC_2H_5$ CH_3 N H CH_3 *II* $\xrightarrow{-H_2}$ NO_2 CH_3CO $COOC_2H_5$ CH_3 N CH_3 *III*

Diagram G 1

Without doubt, reactions of the second group are more frequent, covering a broad range of **enzymatic reactions**. They are very important in biochemistry and, although their applications in organic chemistry are still of little importance, they have economically very favourable production possibilities because of the low energy requirements needed for their realization. The preparation of optically active amino acids and their derivatives from racemic substances by the action of enzymes (2) may serve as an example.

The effect of a catalyst of biological origin, an enzyme, may be illustrated by the reduction of aldehydes and ketones with the help of oxido reductases and a pyridine nucleotide co-enzyme (3). It is now generally known that in the course of reductions and oxidations catalysed by these enzymes, the hydrogen-transfer agent is the pyridine ring of nicotinamide, which is, at the same time, reduced in position 4. A typical coenzyme is nicotinamideadenine dinucleotide ($NAD^{(+)}$), which takes part in the following reaction:

H, $CONH_2$, $N^{(+)}$, R + alcohol $\rightleftharpoons$ H H, $CONH_2$, N, R + aldehyde + $H^{(+)}$

NAD NADH

The stereospecificity of these enzymatic reductions is due to the fact that hydrogen, or possibly deuterium synthetically introduced, transfers only to one side of the nicotinamide ring. In the presence of the same dehydrogenase, hydrogen elimination occurs always from the same side of the dihydronicotinamide ring. The alcoholdehydrogenase (ADH) from yeast has been accepted as the basis for determining the stereospecificity of an enzyme.

Knowledge of the stereospecificity of reduction by means of fermenting yeast, *i.e.* with the help of alcoholdehydrogenase from yeast, was augmented by Prelog *et al.*, who studied fermentative reductions in the bicyclic ketone series. For the reductions they used pure cultures of *Curvularia falcata*, *Aspergillus niger*, and *Streptomyces.* The probable steric course of the reduction of ketones with a decalin arrangement was suggested by Prelog, based on the knowledge of the absolute configuration at the $C_{(4)}$ carbon of the dihydropyridine ring of NADH. Repeated reductions of these ketones (or oxidation of the corresponding alcohols) with the help of *C. falcata* established that the enzyme isolated from these micro-organisms (called *a*-oxidoreductase by Prelog) transfers hydrogen mainly to (or from) the axial position of the substrate, in contrast to the alcohol-dehydrogenase from horse liver, which transfers mainly equatorial hydrogen (*e*-oxidoreductase). In order to explain the course of the reduction and the mutual positions of the substrate and the pyridine ring of the pyridine nucleotide, Prelog used a diamond lattice segment as an aid. The reduction in the presence of alcoholdehydrogenase from yeast and of *C. falcata* enzymes yields alcohols of absolute configuration *S* (Diagram G 2)

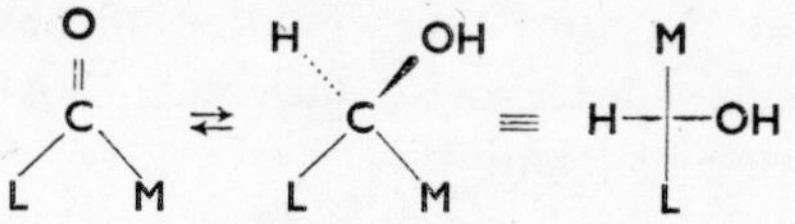

O
‖
C
L M
⇄
H OH
C
L M
═
M
H—|—OH
L

Diagram G 2

The production of alcohol of absolute configuration *S* may be explained if the substrate is coordinated, in the transition state for the reduction, in a plane parallel to the plane of the dihydropyridine ring, such that the large substituent is coordinated in the direction of carbon $C_{(5)}$ and the hydroxyl group in the direction of the nitrogen of the pyridine ring. The smaller substituent (M) is then coordinated in the direction of the amide group.

The reduction of deuterated aldehydes (or the reduction of aldehydes with the help of NADD) also takes place asymmetrically. For example, both enantiomers of ethanol-[1-D] have been prepared in this manner. In the first case, acetaldehyde was reduced with the help of NADD of A-stereospecificity, prepared by means of $NAD^{(+)}$ reduced with ethanol-[1,1-D_2] in the presence of ADH. Deuterated acetaldehyde and NADH was used for the reduction in the second case (Diagram G 3).

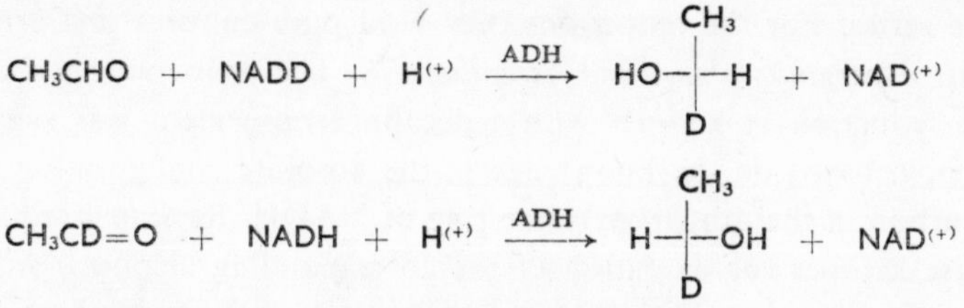

$$CH_3CHO + NADD + H^{(+)} \xrightarrow{ADH} HO-\overset{CH_3}{\underset{D}{|}}-H + NAD^{(+)}$$

$$CH_3CD{=}O + NADH + H^{(+)} \xrightarrow{ADH} H-\overset{CH_3}{\underset{D}{|}}-OH + NAD^{(+)}$$

Diagram G 3

However, from the stereochemical point of view enzymatic asymmetric reactions are highly complicated and only in rare cases are we able to express the reaction course as in the stated example. Their more detailed discussion is outside the scope of this book. We will therefore point out at least the most important characteristics of enzymatic reactions from the aspect of stereochemistry: a) a high degree of stereoselectivity; b) an extreme sensitivity towards steric effects; c) the stereoselective nature of some reactions in which we would not normally expect to observe stereoselectivity.

From the aspect of organic chemistry, the most interesting reactions are those in which the asymmetric component is the reacting substance, the reagent molecule, or the molecule of a catalyst or solvent present in the reaction medium.

The course of an asymmetric reaction is, as a rule, controlled by kinetic factors, which means that the relative proportions of the products in the mixture are determined by the rate of formation of the individual isomers. In kinetically controlled reactions, the chiral structure of the components in the first place influences the access of the reagent to the molecule of the reacting substance. The ease of access from different sides to the bond at which the reaction occurs differs because of the asymmetric substitution in the vicinity of the bond. The diastereoisomeric transition states through which the reactions proceed are characterized by different free energy contents. The reaction *via* a transition state with a higher free energy is of course less probable and less frequent. Most of the molecules of the final product will be sterically arranged in agreement with the lower-energy transition state. The differences in the free energy contents of both transition states depend on the nonbonded interactions between the groups of atoms of the starting material and the particles of the newly added reagent. They are thus influenced in the first place by the flexibility of the chain, which may improve the accessibility of a reactive centre otherwise sterically hindered, and by the effective steric volume of the substituents. More complex organic cyclic molecules, accommodating *e.g.* several asymmetric carbons, or compounds with multiple bonds are, as a rule, sufficiently rigid to allow the access of the reagent from one side only. Reactions with them lead predominantly to one of the possible epimers. In the case of aliphatic compounds, the influence in most cases is limited to the asymmetric carbon nearest to the reacting part of the molecule because the essentially free rotation of the chain allows the more distant substituents to adopt a position such that they do not have to interfere with the entering reagent. The product of the reaction in these cases usually is a mixture of both possible epimers, with one of them predominating. The geometrical arrangement of the transition state* is, however, generally predictable, and therefore it is usually possible to derive empirical relationships between the steric structure of the starting material and the predominating reaction product.

Asymmetric Reactions Conditioned by the Presence of an Asymmetric Centre in the Reacting Substance

A typical example of asymmetric reactions of this type is the reaction of an equimolar amount of a Grignard reagent with the ester of an α-keto acid and an optically active alcohol. The keto group, which is considerably more reactive towards organomagnesium compounds than the esterified carboxyl, is converted into a tertiary carbinol, whereas the ester group remains unchanged. The final product of the reaction after saponification is an optically active tertiary α-hydroxy acid (Diagram G 4). The preferred formation of one of the enantiomers is caused by

$$R_1{-}CO{-}COOC_1(R_2)(R_3)(R_4) + R_5MgX \longrightarrow$$

$$\longrightarrow R_1{-}C(R_5)(OH){-}COOC_1(R_2)(R_3)(R_4) \xrightarrow{H_2O} R_1{-}C(R_5)(OH){-}COOH$$

Diagram G 4

* Sufficient knowledge of the reaction mechanism is an apparent pre-condition which, however, is not always fulfilled.

asymmetric induction, conditioned by the different thermodynamic stabilities of the transition states formed by coordination of the organomagnesium halide to opposite sides of the planar arrangement of the bonds at the carbonyl carbon atom. This idea was worked out in detail by Prelog (4) in 1950, who with its help not only characterized the mechanism of asymmetric carbon atom generation but also derived the configurational relationships between the original alcohol and the α-hydroxy acid produced.

Let us imagine the ester molecule as an arrangement in which the grouping of atoms C—CO—CO—O—$C_{(1)}$— is coplanar with the bonds exclusively in *anti*-periplanar positions (spectroscopic measurements show that this is in fact true) and the most bulky substituent at $C_{(1)}$ also in an *anti*-periplanar position (see Diagram G 5). We then see that the ease of access of the nucleophilic reagent to the carbon of the carbonyl group differs according to whether the reagent approaches from the side of the small substituent S (with the least effective steric volume) or from the side of the medium substituent M. In the latter case, the transition

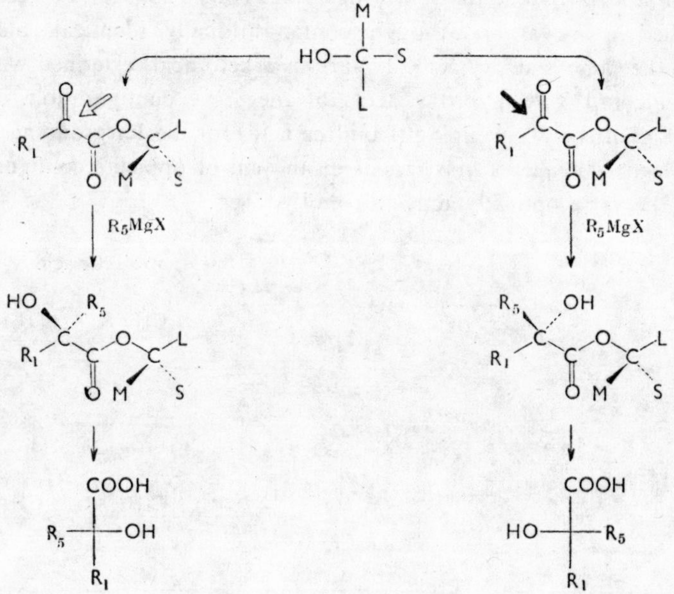

Diagram G 5

state formed will have a higher free energy content because of the larger interaction between substituent M and the approaching reagent. Therefore, of the two possible diastereoisomeric hydroxy acid esters, a larger quantity of the first one will be formed and its prevalence will be manifested by the optical activity of the isolated acid.

Diagram G 6 then shows that the esters of α-keto acids with alcohols corresponding to Formula *IV* lead to hydroxy acids (*VI*), whereas esters with an opposite configuration of the auxiliary optically active alcohol *V* lead to their enantiomers (*VII*). In all cases when the substituents at the asymmetric atom differ sufficiently in bulk to enable us to unequivocally ascribe the symbols S, M and L and to assume the substantially easier access of the reagent from the side of substituent S, we are offered the possibility of predetermining the probable configuration of the hydroxy acid. Since the absolute configuration of these acids is generally known or easy to determine, it is possible to establish the absolute configuration of the applied optically active alcohol from the sign of rotation of the mixture of hydroxy acids. On the basis of such considerations we are then able to arrive at the following useful conclusions . 1. A keto acid esterified with various, though configurationally identical, alcohols yields the same hydroxy acid. 2. Various α-keto acids esterified with the same alcohol give hydroxy acids of the same configuration. 3. An exchange of the R_1 grouping (from the acid) for the R_5 grouping (from the Grignard reagent) must result in an acid of opposite configuration when the same optically active alcohol is used.

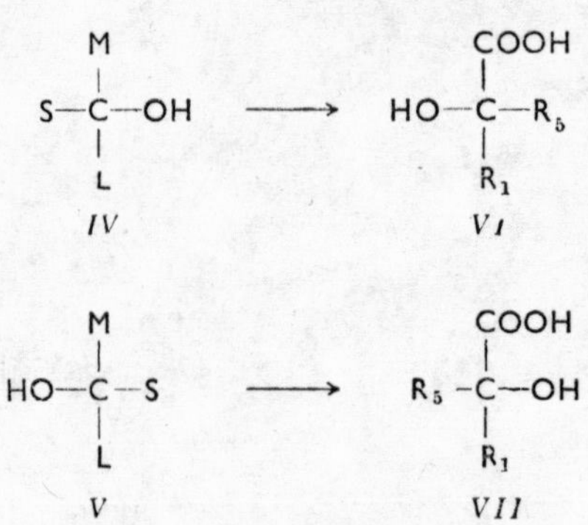

```
      M                      COOH     L—(−)-atrolactic acid
      |                       |
   S—C—OH      ——→      HO—C—R5
      |                       |        R1 = CH, R5 = (CH3)
      L                       R1
     IV                       VI

      M                      COOH     D—(+)-atrolactic acid
      |                       |
  HO—C—S       ——→      R5—C—OH
      |                       |        R1 = C6H5, R5 = CH3
      L                       R1
      V                      VII
```

Diagram G 6

The reaction of organomagnesium reagents with α-keto esters (especially with phenylglyoxylic esters) may be considered a general method for the determination of the configuration of asymmetric carbinols. By this method, Prelog *et al.* established for example the absolute configuration of steroid hydroxy-derivatives and thus also of the whole steroid molecule, in agreement with the independently executed determination by means of three-dimensional X-ray analysis. It has also been successfully proved that pentacyclic triterpenes have the same absolute configuration as the respective part of the steroid molecule. This discovery was of great significance for the elucidation of the genetic relationships between plant products of both types and has been lately verified by comparing the optical rotatory dispersion curves. In principle, this may be extended to include also the determination of the absolute configuration of amines, because the carbonyl group of the ketone type reacts preferably with the Grignard reagent in the amide of an α-keto acid as well.

Until now we have considered only the qualitative relationships between the optical activity of the starting alcohols and the products obtained. However, in addition to the sign of rotation, the ratio of the optically active isomers formed is interesting. In most cases (provided a product of one structural type is formed), this ratio is easily assessable by comparing the rotation of the mixture formed by the reaction with the rotation of the pure isomer. The surplus percentage of one enantiomer was termed by Prelog the **optical yield of the reaction**. Without doubt, the larger the differences in the effective volume of the three substituents at the asymmetric carbon of the carbinol, the larger will be the difference of the rate of formation of both possible diastereoisomeric hydroxy esters, and therefore the larger will be the optical yield of the reaction. By the correct choice of the auxiliary alcohol it is possible to increase the optical yield of the reaction. Secondary methyl aryl carbinols with extremely bulky aryl groups proved to be most efficient, especially 1-(2,4,6-tricyclohexylphenyl)ethanol (see Table XII). The optical yield of reactions with secondary cyclic alcohols (*e.g.* borneol) is relatively poor, and because substances of this type generally include several asymmetric atoms the results are difficult to interpret from the aspect of absolute configuration. Their total steric action is of course due to the overall shape

Table XII

THE OPTICAL YIELD OF REACTIONS OF METHYLMAGNESIUM BROMIDE WITH PHENYLGLYOXYLIC ESTERS OF SEVERAL OPTICALLY ACTIVE CARBINOLS

Alcohol	*Optical yield,*[a] %	*Alcohol*	*Optical yield,* %
1-Phenylethanol	3	(−)-menthol	25
1-α-Naphthylethanol	12	(+)-neomenthol	12
1,1,1,-Triphenyl-2-propanol	49	(+)-borneol	11
1-(2,4,6-Trimethylphenyl)ethanol	30	(−)-isoborneol	8
1-(2,4,6-Tricyclohexylphenyl)ethanol	66	—	—

[a] The total yield of the mixture of both atrolactic acid enantiomers was 70−80 % of the original ester.

of the molecule and not to the surroundings of a single asymmetric carbon atom.

The role of the asymmetric component of the reaction mixture may of course be played also by substances the optical activity of which is due to chirality elements other than the asymmetric carbon atom. Atropisomers of the biphenyl type have been studied experimentally from this point of view, and generally have resulted in a large optical yield (see Table XIII). At the same time, the reactions enable us to establish the relationship between the steric arrangement of the atropisomer molecule and the derivative with the asymmetric carbon atom.

The esters of α-keto acids react with other nucleophilic reagents, for example lithium aluminium hydride or aluminium amalgam, in a similar manner to their reactions with organomagnesium compounds. Reduction by means of amalgam also gives α-hydroxy acids; however, in contrast to the reaction with the Grignard reagents, the hydroxyl group is secondary. The hydride always attacks both functions capable of reduction, the ketone as well as ester grouping, and leads exclusively to glycols with one primary and one secondary hydroxyl. Since the absolute configuration of both types of products is well known, these reactions may also be applied to the determination of absolute configuration. However, the optical yield drops in the series $a > b > c$ (see Diagram G 7).

Table XIII

THE OPTICAL YIELD OF D(+)-ATROLACTIC ACID FROM THE REACTION OF METHYLMAGNESIUM IODIDE WITH PHENYLGLYOXYLIC ESTERS OF CERTAIN ATROPISOMERS

Substance	*Optical yield,* %
VIII	70
IX, $R_2 = CH{=}CH_2$; $R_3 = CH_2CH(C_6H_5){-}N(CH_3)_2$	91
IX, $R_2 = CH_2CH_3$; $R_3 = CH_2CH(C_6H_5){-}N(CH_3)_2$	89
IX, $R_2 = CH{=}CH_2$; $R_3 = CH{=}CH{-}C_6H_5$	93

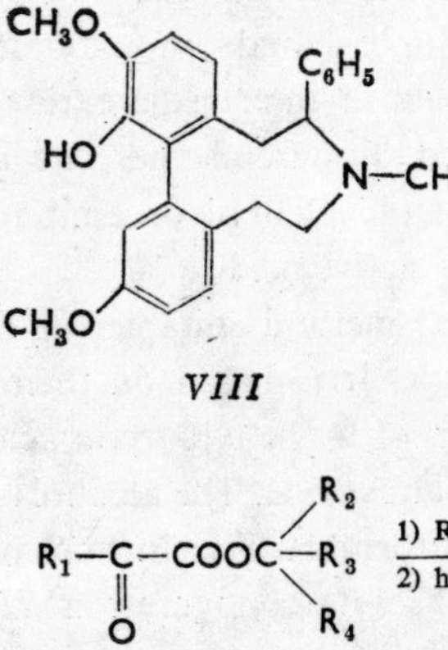

VIII

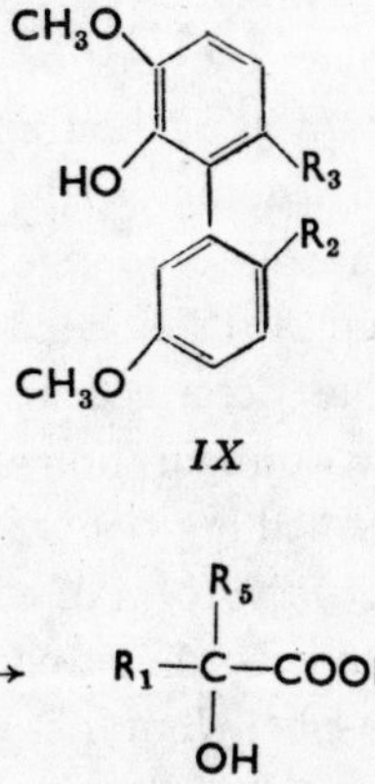

IX

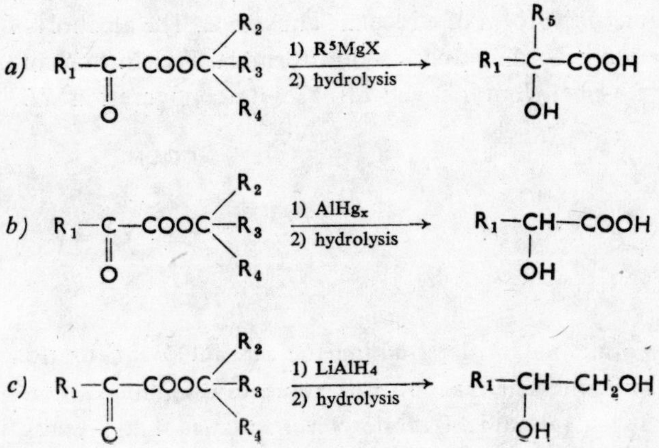

a) $$R_1{-}CO{-}COOC(R_2)(R_3)(R_4) \xrightarrow[\text{2) hydrolysis}]{\text{1) } R^5MgX} R_1{-}C(R_5)(OH){-}COOH$$

b) $$R_1{-}CO{-}COOC(R_2)(R_3)(R_4) \xrightarrow[\text{2) hydrolysis}]{\text{1) } AlHg_x} R_1{-}CH(OH){-}COOH$$

c) $$R_1{-}CO{-}COOC(R_2)(R_3)(R_4) \xrightarrow[\text{2) hydrolysis}]{\text{1) } LiAlH_4} R_1{-}CH(OH){-}CH_2OH$$

Diagram G 7

The removal of the optically active component further from the reaction centre has the effect of reducing the optical yield of the reaction because the differences in the stabilities of the respective transition states are thus also reduced, especially in the case of mobile aliphatic molecules. This is the case with the reactions in the α-, γ- and δ-keto acid series (with β-keto acids the reaction cannot be carried out). Of course, no conclusions concerning the relationship of the configuration of the starting material to the configuration of the product may be drawn from these reactions because the conformation of the starting material is uncertain.

Attention has lately been given to the use of asymmetric transformation for the determination of the absolute configuration of organic compounds. The reactions of racemic compounds (mainly acids with an asymmetric carbon in α-position to the carboxyl group) with optically active alcohols or amino compounds, which have their hydroxyl or amino group attached to the asymmetric carbon, yield the two possible diastereoisomeric esters or amides in unequal amounts. If we know the conformation of the transition state or intermediate, respectively, or the reaction leading to the preferred diastereoisomer, we are able to assign the absolute configuration of the alcohol or amino compound from the sign of the isolated optically active acid.

Horeau has described an empirical method suitable for determining the absolute configuration of alcohols. It is based on their being preferably esterified by one enantiomer of α-phenylbutyric acid, used for the reaction in the form of a racemic anhydride. The alcohols of absolute configuration according to projection formula *X* yield unchanged optically active α-phenylbutyric acid of absolute configuration *XI*.

L
HO—|—H
M

X

COOH
H—|—C_2H_5
C_6H_5

XI

Mislow's method of determining the absolute configuration of alcohols is based on their reaction with toluene-*p*-sulphinyl chloride in pyridine, giving a mixture of diastereoisomeric sulphinyl esters which is then converted into optically active methyl *p*-tolyl sulphoxide by the action of methylmagnesium iodide. Alcohols of absolute configuration *XII* yield a surplus of laevorotatory sulphoxide of absolute configuration *S* (*XIII*).

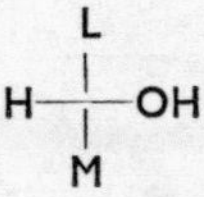

$$\begin{array}{ccc} & L & \\ & | & \\ H - & + & - OH \\ & | & \\ & M & \end{array}$$

XII

$$\begin{array}{c} S-C_6H_4-CH_3 \\ O \quad CH_3 \end{array}$$

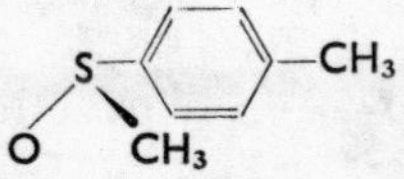

XIII

The determination of the absolute configuration of sulphoxides is based on a similar principle.

Červinka made use of asymmetric transformations in order to determine the absolute configuration of amino compounds and also of acids. The reaction is based on the acylation of compounds containing an amino group at the asymmetric carbon by means of racemic hydratropic acid in the presence of dicyclohexylcarbodi-imide. We may assign the absolute configuration of the carbon atom carrying the amino group from the sign of rotation of the unchanged acid. The asymmetric course of the reaction of hydratropic acid with optically active amino compound may be explained by the formation of two transition states differing in energy. A larger amount of one of the two possible diastereoisomeric amides is thus formed. The configuration of the isolated unchanged optically active hydratropic acid is opposite to that of the acid taking part in the formation of the preferred diastereoisomeric amide. Consequently it may be said that secondary amines, α-amino alcohols (L = —CH(OH)R), and esters of α-amino acids (L = COOR) of the same configuration (*XIV*) will lead to *S*-(+)-hydratropic acid (*XV*):

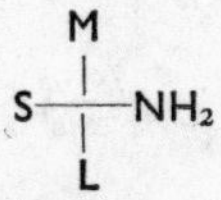

$$\begin{array}{ccc} & M & \\ & | & \\ S - & + & - NH_2 \\ & | & \\ & L & \end{array}$$

XIV

$$\begin{array}{ccc} & COOH & \\ & | & \\ CH_3 - & + & - H \\ & | & \\ & C_6H_5 & \end{array}$$

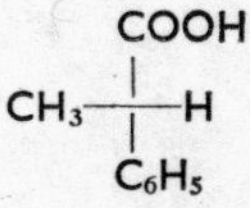

XV

This method proved capable of modification as needed; optically active hydratropic acid may be used for the reaction with racemic compounds and the reaction described may just as well be utilized for the determination of configuration of acids with an asymmetric carbon atom attached in α-position to the carboxyl.

The reaction of ethyl esters of L-α-amino acids with surplus DL-benzyloxycarbonyl-α-amino compounds in the presence of dicyclohexylcarbodi-imide leads to optically active L-benzyloxycarbonyl-α-amino acids.

Asymmetric Reformatsky Reaction

Unlike the reaction of α-keto esters, in this case an asymmetric component, represented by the bromoacetate of an optically active auxiliary alcohol, is added to the carbonyl group of a ketone which does not include any chirality element. The course of the reaction of acetophenone with (−)-menthyl bromoacetate, yielding dextrorotatory β-phenyl-β-hydroxybutyric acid, is again due to preferred reaction *via* one of two transition states of unequal energy.

Asymmetric Addition to Double Bond between Two Carbon Atoms

The catalytic hydrogenation of esters of *trans*-β-methylcinnamic acid with optically active alcohols and subsequent hydrolytic decomposition may be used in order to prepare optically active β-phenylbutyric acid. If we assume that the hydrogenated substance approaches the planar surface of the catalyst from the side of the smallest substituent at the asymmetric carbon, we may derive the relationship between the configuration of the original alcohol and the acid produced (see Diagram G 8):

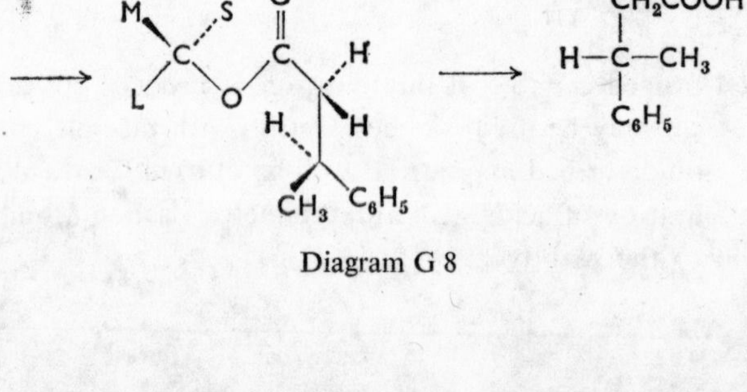

Diagram G 8

The hydroxylation of the double bond, by the action of potassium permanganate, of the ester of fumaric acid and (−)-menthol, yields, after hydrolytic decomposition, laevorotatory tartaric acid. Here also we may formulate a configurational relationship between the starting material and the product (see Diagram G 9). The interpretation of both reactions

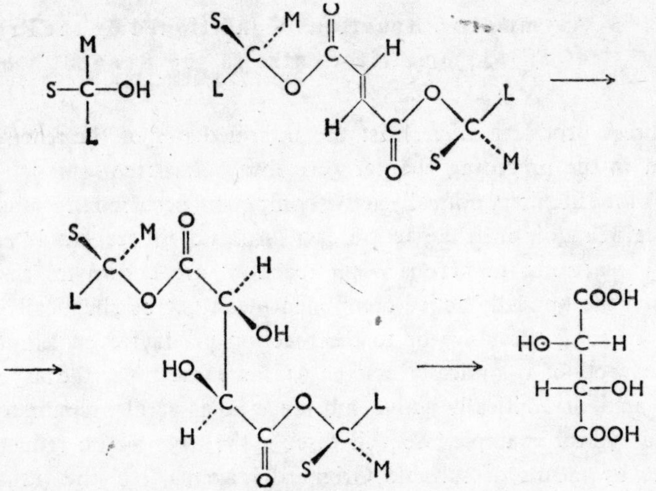

M
S—C—OH
L

S M O
C C H
L O
H C O L
C
O S M
→

S M O
C C H
L O C
OH
→ HO →
C O L
H C C
O S M

COOH
HO—C—H
H—C—OH
COOH

Diagram G 9

referred to is of course impeded by the uncertainty regarding the actual conformation of the ester at the instant of addition. The conformations used in Diagrams G 8 and G 9 follow from reactions with substances of known absolute configuration.

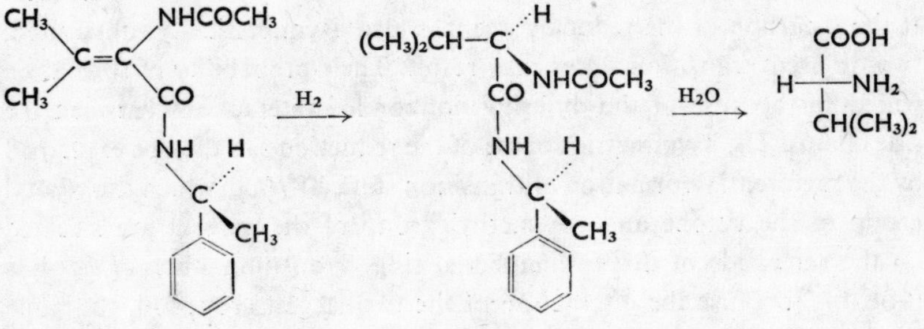

CH_3 NHCOCH_3
C=C
CH_3 CO
NH H
C
CH_3

$\xrightarrow{H_2}$

H
$(CH_3)_2CH$—C
CO NHCOCH_3
NH H
C
CH_3

$\xrightarrow{H_2O}$

COOH
H—NH$_2$
CH$(CH_3)_2$

Diagram G 10

The catalytic hydrogenation of the (−)-α-phenylethylamide of α-acetamido-β,β-dimethylacrylic acid on Raney nickel leads to a saturated amide, the hydrolysis of which yields D-valine of 39 % optical purity (Diagram G 10)

Asymmetric Reactions Conditioned by the Presence of an Asymmetric Centre in the Reagent Molecule

The direct products or at least the intermediates of the reactions described in the preceding chapter were always diastereoisomeric. The removal of the auxiliary optically active component occurred in a subsequent chemical reaction or by the action of a surplus of the reagent. In addition to such reactions, numerous other reactions are known in the course of which the optically active component need not be chemically bound to the reacting substance or to the reaction product even temporarily, yet the product is optically active. At the same time, the asymmetry of the auxiliary optically active substance is, as a rule, eliminated.

As a typical example, we will discuss the asymmetric reduction of ketones by means of suitable Grignard reagents, *e.g.* the reaction of t-butyl methyl ketone (*XVI*) with (+)-2-methylbutylmagnesium chloride. The secondary alcohol formed, 3,3-dimethyl-2-butanol (*XVII*), is optically active. It is known that Grignard reagents act as reducing agents by means of a cyclic mechanism *via* a complex in which the magnesium atom is co-ordinated with the oxygen atom of the carbonyl, and the hydrogen atom at the β-carbon of the reagent is utilized for the reduction proper. The reduction proceeds most smoothly if the β-carbon is tertiary. If the β-carbon of the reducing agent is also asymmetrically substituted, two different transition states may result. Their probability of formation will differ because of the differing nonbonded interactions between the side chains. The asymmetric course of the reduction can then be explained by the preferential formation of transition state *XVIII*, in which the t-butyl group of the ketone and the methyl group of the reagent are situated on the same side of the six-membered ring. Transition state *XIX* is less probable, because the interaction of the two largest groups on the same side of the ring is much greater than in the first case. It is again possible to formulate a relationship between the absolute configuration of the alkyl halide and that of the alcohol formed. If the branching of the alkyl halide chain, which causes its optical activity, is at the γ-position, the reduction product is formed to a small extent only and the optical yields are low. This agrees with the formulation of transition state *XX*. Here also, hydrogen transfer occurs from the β-carbon, and both hydrogen atoms of the methylene group in this position may enter the reaction with similar probability (Diagram G 11) (5).

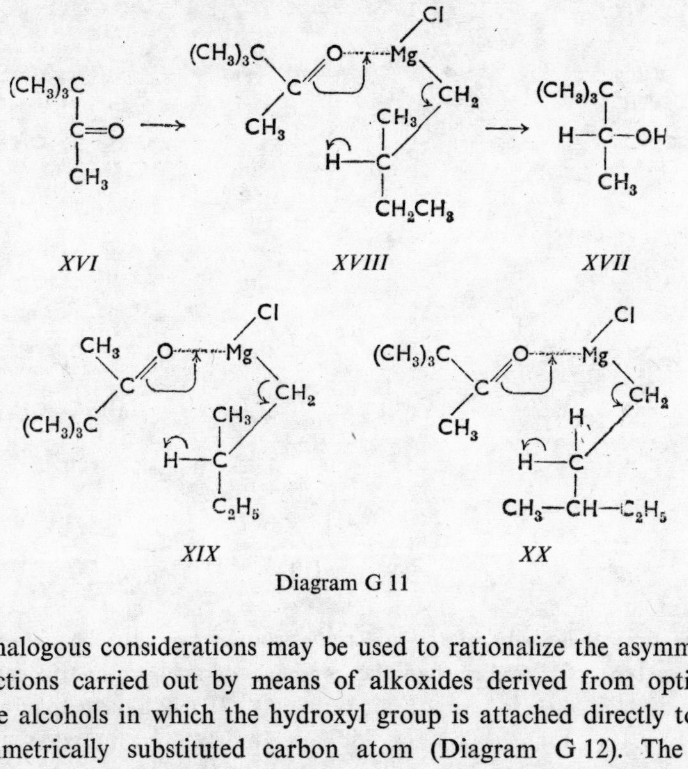

XVI *XVIII* *XVII*

XIX *XX*

Diagram G 11

Analogous considerations may be used to rationalize the asymmetric reductions carried out by means of alkoxides derived from optically active alcohols in which the hydroxyl group is attached directly to the asymmetrically substituted carbon atom (Diagram G 12). The two transition states possible during the **Meerwein-Ponndorf** reduction with aluminium alkoxides* differ from each again by the ease of their forma-

* We have to bear in mind that the Meerwein-Ponndorf reduction is reversible; the considerations therefore hold only for reactions proceeding for a sufficiently short period (several hours at most) before equilibrium is established.

tion. A lower activation energy is sufficient in order to achieve transition state *XXI* than is necessary for state *XXII*; the former will therefore predominate and determine the configuration of the predominant enantiomer *XXIII*. In the case of the homologous alkoxide with the asymmetric carbon atom in the α-position to the hydroxyl group, a cyclic transition state cannot be formed with the asymmetrically substituted atom as a component of the ring. The attempt to perform asymmetric reduction with the help of (−)-aluminium 2-methylbutoxide failed but is possible in principle.

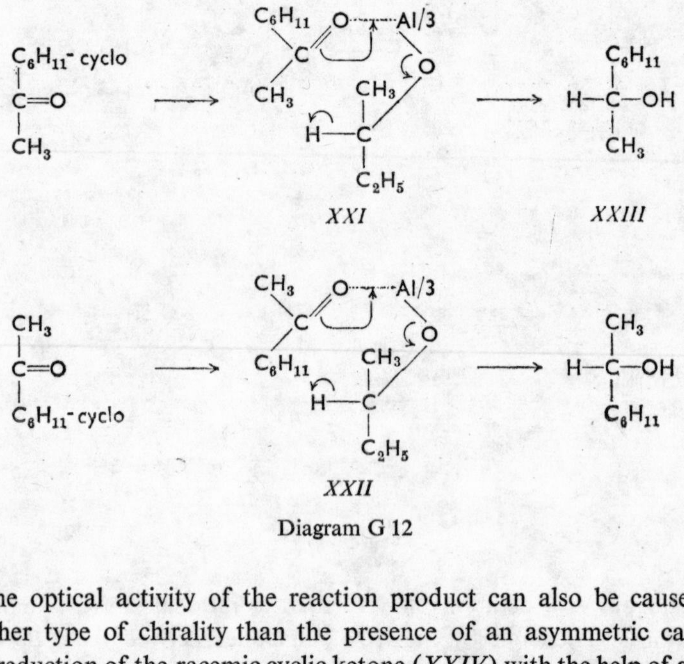

C_6H_{11}-cyclo–C(=O)–CH_3 ⟶ *XXI* ⟶ H–C(C_6H_{11})(CH_3)–OH

XXI　　*XXIII*

CH_3–C(=O)–C_6H_{11}-cyclo ⟶ *XXII* ⟶ H–C(CH_3)(C_6H_{11})–OH

XXII

Diagram G 12

The optical activity of the reaction product can also be caused by another type of chirality than the presence of an asymmetric carbon. The reduction of the racemic cyclic ketone (*XXIV*) with the help of a mixture of (+)-3,3-dimethyl-2-butanol and aluminium t-butoxide yields, in the case of one enantiomer, the cyclic carbinol (*XXV*), whereas the second enantiomer accumulates unchanged. From the study of molecular models* of the stated substances it follows (Diagram G 13) that the

* In Formula *XXVIA*, both benzene rings are approximately at right angles to each other, and are situated in such a manner that if we imagine the molecule as

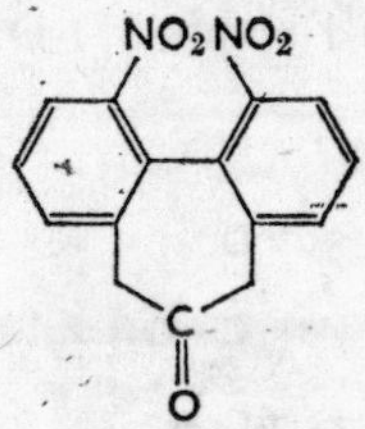

XXIV 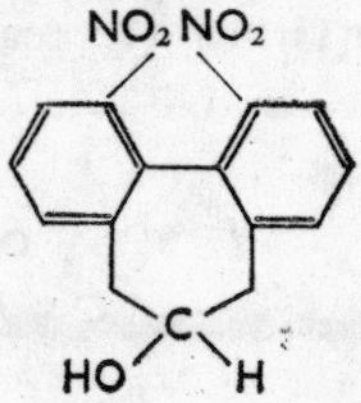 *XXV*

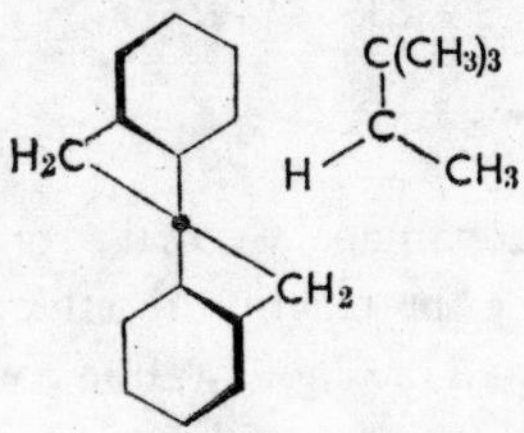

XXVIA 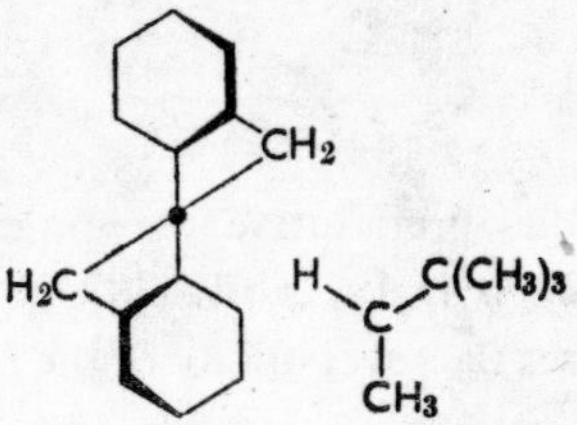 *XXVIB*

Diagram G 13

carbonyl group of atropisomer *XXVIB* is considerably more accessible for the molecule of the reducing agent, because no interaction of the bulky t-butyl group with the aromatic ring takes place and only the small methyl group approaches the ring. This isomer is therefore more easy to reduce and will consequently predominate in the alcohol fraction. On the basis of such ideas we may also formulate the spatial relationships between atropisomers of the biphenyl type and substances with asymmetric atoms.

The magnesium alkoxides formed by the decomposition of Grignard reagents by means of asymmetric alcohols may also be applied to analogous asymmetric reductions. As an example we may include the reaction of the alkoxide prepared from (−)-2-octanol-[2-D] and an equimolar quantity of ethylmagnesium bromide with butyraldehyde

---

a propeller rotating around the axis perpendicular to the plane of the paper in the clockwise sense the whole configuration moves away from the observer; under the same conditions in the case of Formula *XXVIB* it moves towards the observer. In both formulas the hydroxyl group of the alcohol protrudes in front of the plane of the paper.

(Diagram G 14). The reaction product is (+)-1-butanol-[1-D] (*XXVII*) with a measurable rotation:

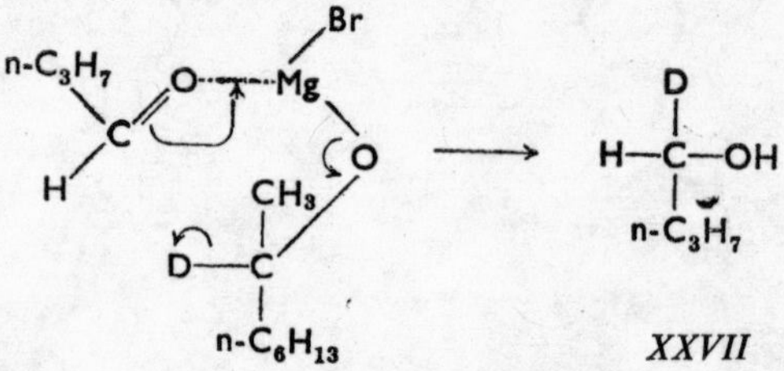

XXVII

Diagram G 14

From the preparative viewpoint, the advantage over the preceding procedure is in the possibility of executing the reaction in ether, which suppresses the reversibility of the reaction and the equilibration connected with it.

A special case of asymmetric reactions is constituted by reductions with the help of optically active alkoxylithium aluminium hydrides, *i.e.* reagents prepared by the reaction of lithium aluminium hydride with one equivalent of an optically active alcohol (*e.g.* (−)-menthol), according to the equation

$$LiAlH_4 + ROH \rightarrow LiAlH_3(OR)$$

The reduction of 1-methyl-2-alkyl-Δ^2-piperidine salts and 1-methyl-2-alkyl-Δ^2-pyrroline salts yields such partially optically active 1-methyl-2-alkylpiperidines (*XXVIII*) and pyrrolidines (*XXIX*).

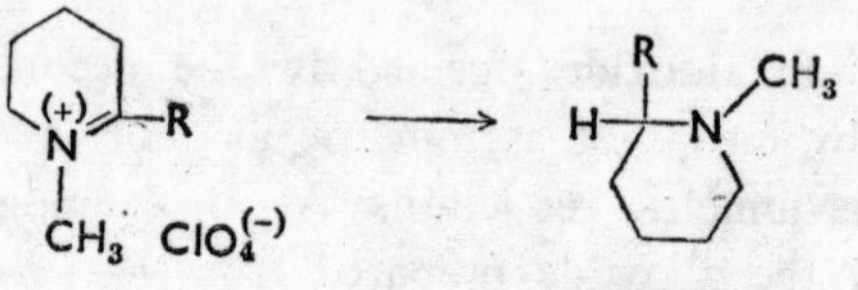

XXVIII

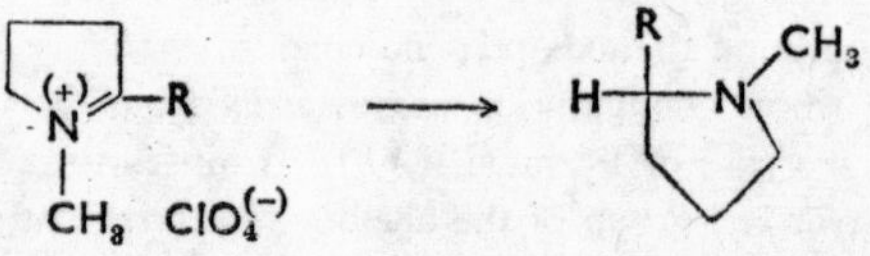

XXIX

Asymmetric reductions of ketones may be carried out in a similar manner. Experiments with $LiAlD_4$ showed that the reduction is really executed by the hydride hydrogens and that a modification of the Meerwein-Ponndorf reaction by the formed alkoxide is not the case. The relationship between the absolute configurations of the initial alcohol and of the product of reduction was in this case determined with the help of cinchona alkaloids as the starting optically active alcohols (see Table XIV). The reduction of methyl alkyl ketones and methyl aryl ketones in the presence of (−)-quinine and (−)-cinchonidine gave the dextrorotatory carbinols *XXXA* and *XXXB*. Conversely, in the presence of dextrorotatory quinidine and cinchonine alkaloids, which are epimers of the first pair at the $C_{(8)}$ and $C_{(9)}$ atoms, the laevorotatory carbinols *XXXC* and *XXXD* (Diagram G 15) are preferably formed. Insufficient knowledge of the arrangement of the lithium aluminium hydride molecules in solution prevents the formulation of probable transition states. However, empirical findings could have been used for the determination of the hitherto unknown absolute configuration of diaryl carbinols and cyclohexyl aryl carbinols. The scope of the reaction was broadened to include the reduction of imines as well (6).

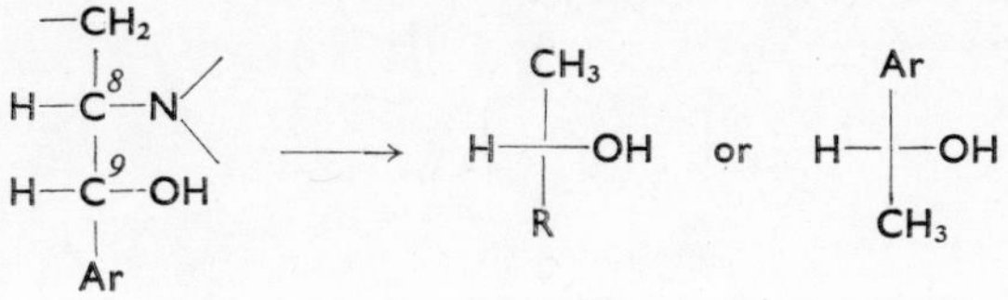

(−)-quinine
(−)-cinchonidine

XXXA *XXXB*

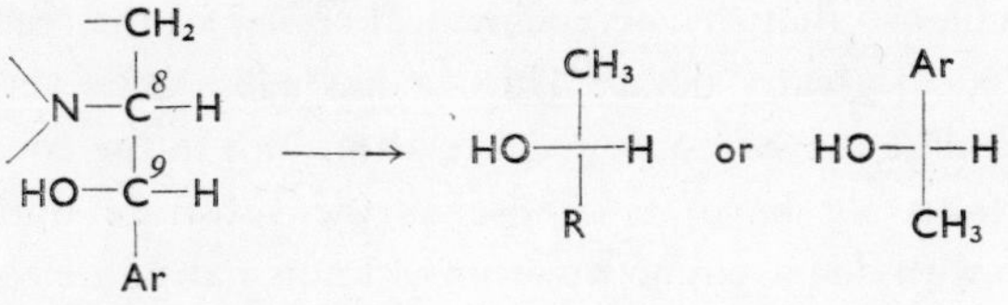

(+)-quinidine
(+)-cinchonine

XXXC *XXXD*

Diagram G 15

Asymmetric Reactions Conditioned by Participation of an Optically Active Catalyst

Very few asymmetric reactions brought about by the presence of a simple chemical catalyst are known. There is only one significant example, the asymmetric synthesis of cyanohydrins, catalysed by the presence of optically active bases – alkaloids. Under these conditions, the product of the reaction of an aldehyde with hydrogen cyanide is an optically active α-hydroxy nitrile, more or less racemized according to the character of the catalyst and the reaction time.

The experimentally investigated addition of hydrogen cyanide to cinnamaldehyde (*XXXI*), catalysed by cinchona bark alkaloids (*XXXII*), always leads to an optically active product (see Table XIV). The optical yield is influenced principally by the configuration at the $C_{(9)}$ atom;

$$C_6H_5CH{=}CHCHO + HCN \longrightarrow C_6H_5CH{=}CH{-}CH(OH)(CN)$$

XXXI

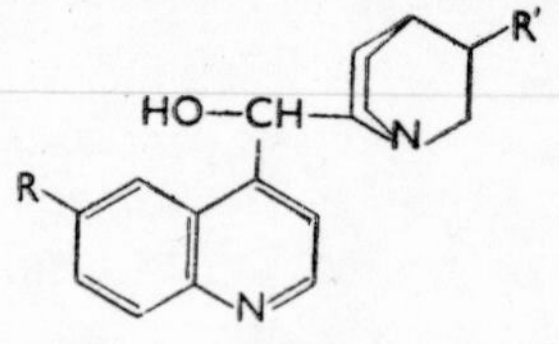

XXXII

however, a significant role will also be played by the other asymmetric carbons in the molecule of the catalyst as well as by the overall shape of the molecule, so that the stereochemical course of the reaction cannot be interpreted in greater detail. Thus, it has again been verified that in order to elucidate the stereochemical relationships in the course of asymmetric reactions we have to choose as the substrate optically active components with one asymmetric atom of known absolute configuration. A reaction that is worthy of attention is the addition of (–)-di-isopinocampheylborane, obtained by the hydroboration of α-pinene, to an olefin double bond. Optically active alcohols of high optical purity are formed

Table XIV

RESULTS OF REACTION OF HYDROGEN CYANIDE WITH CINNAMALDEHYDE, CATALYSED BY CINCHONA BARK ALKALOIDS

Substituent at $C_{(6)}$ XXXII	*Alkaloid*	*Configuration at*		*Optical yield, %*	*Acid configuration*
		$C_{(8)}$	$C_{(9)}$		
OCH_3	quinidine	D	L	5.2	L
OCH_3	quinine	L	D	5.7	D
OCH_3	9-epiquinidine	D	D	2.6	D
OCH_3	9-epiquinine	L	L	2.4	L
H	cinchonine	D	L	1.2	L
H	cinchonidine	L	D	1.1	D

after oxidation. Pracejus has described the reaction of unsymmetrically substituted ketenes with alcohols in the presence of optically active alkaloids, producing partially optically active esters.

Asymmetric Reactions Conditioned by an Optically Active Solvent

The overwhelming majority of the large number of attempts by various authors to carry out an asymmetric synthesis with the help of an optically active solvent were wrong in principle and therefore unsuccessful. As an example, we may cite the attempt at an asymmetric reduction of phenylglyoxylic acid with the help of zinc, in an aqueous solution of tartaric acid. The reaction is caused by the hydrogen and tartaric acid does not participate at all in the addition of hydrogen to the C=O double bond. Success may be expected only in the case of such reactions, in which the molecules of the optically active solvent solvate the starting material or reagent in a defined and stable manner.

The use of optically active ethers in addition reactions of Grignard reagents may serve as an example. It is known that two molecules of an ether or tertiary amine, which are used as solvents, are always attached by co-ordination to the organomagnesium halide molecule and thus form an important component of the reagent. Prior to the addition proper of the reagent, one molecule of the solvent in the complex is displaced by a molecule of the reacting substance. The presence of the second solvent molecule then permits two different arrangements when passing to the transition state. After many unsuccessful experiments, an asymmetric reaction in (+)-2,3-dimethoxybutane succeeded, first with a complicated reagent prepared from 2,2,6-trimethyl-6-chlorocyclohexanone (*XXXIII*) and later on even with simple alkylmagnesium·halides. The

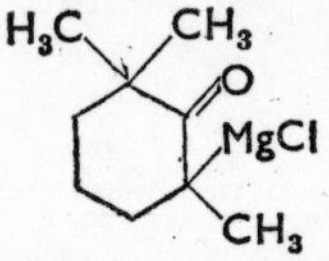

H_3C, CH_3, O, MgCl, CH_3

XXXIII

dependence of the optical yield on the nature of the reagent halogen is in full agreement with the stated concept. The optical activity for the reaction of benzaldehyde with an ethylmagnesium halide (Diagram G 16)

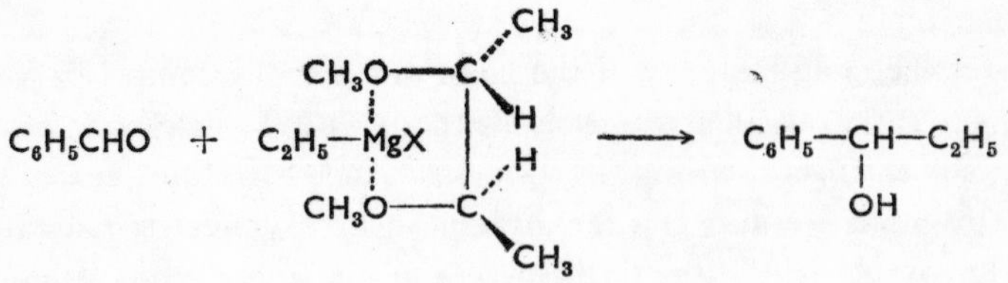

$$C_6H_5CHO + C_2H_5{-}MgX \cdot (CH_3O{-}CH(CH_3){-}CH(CH_3){-}OCH_3) \longrightarrow C_6H_5{-}CH(OH){-}C_2H_5$$

Diagram G 16

decreases in the order Br > 1 > Cl. The capability of ethylmagnesium halides to form complexes with two molecules of ether decreases in the same order; the formation of the complex is already very improbable in the case of C_2H_5MgCl. The presence of two ether groups in the solvent molecule is of fundamental importance for the asymmetric course of the reaction. Reagents prepared in optically active solvents accommodating only one ether group yielded optically inactive products only.

Literature

1. Kuhn W., Knopf E.: Z. Phys. Chem. *7*, 292 (1930); Chem. Zentr. *1930*, I, 3529.
2. Greenstein J. P., Winitz M.: *Chemistry of the Amino Acids.* Vol. 1, p. 728. Wiley, New York, 1961.
3. Červinka O., Hub L.: Chem. listy *60*, 34 (1966).
4. Prelog V.: Bull. soc. chim. France *1956*, 987.
5. Mosher H. S., Stevenot J. E., Kimble D. O.: J. Am. Chem. Soc. *78*, 4374 (1956).
6. Červinka O.: Collection Czechoslov. Chem. Commun. *26*, 673 (1961) *30*, 1684, 1693, 1738, 2484, 2487, 3462 (1965).

LITERATURE RECOMMENDED

1. Eliel E. L.: *Stereochemistry of Carbon Compounds*. McGraw-Hill, New York, 1962.
2. Freudenberg K.: *Die Stereochemie*. Deuticke, Leipzig, 1932.
3. Klabunovskij E. I.: *Asimmetricheskij sintez*. Moscow, 1960.
4. Kuhn W.: *Possible Relation between Optical Activity and Aging*. Advances in Enzymology *20*, 1 (1958).
5. Neuberger A.: *Stereochemistry of Amino Acids. Advances in Protein Chemistry*, Vol. 4, p. 297. Academic Press, New York, 1948.
6. Newman M. S.: *Steric Effects in Organic Chemistry*. Wiley, New York, 1956.
7. Pauling L., Corey R. B.: *The Configuration of Polypeptide Chains in Proteins. Fortschritte der Chemie Organischer Naturstoffe*, p. 110. Springer, Vienna, 1954.
8. *Progress in Stereochemistry* (W. Klyne and P.B.D. de la Mare, Eds.). Butterworths, London, Vol. 1, 1954, Vol. 2, 1958, Vol. 3, 1962.
9. Temnikova T. O.: *Kurs teoreticheskich osnov organicheskoj chimii*. Moscow, 1962.
10. Eliel E. L., Allinger N. L., Angyal S. J., Morrison G. A.: *Conformational Analysis*. Wiley, New York, 1965.
11. Mislow K.: *Introduction to Stereochemistry*. Benjamin Inc. New York, 1966.
12. Eliel E. L., Allinger N. L.: *Topics in Stereochemistry*. Vol. 1—4, Wiley, New York, 1967—1969.

INDEX